"This important and stimulating book argues that we are capable of bouncing ourselves out of locked-in ways of thinking, ways of thinking unhelpful at best and catastrophic at worst. The book's bold ambition gives us the concepts and language to navigate the perilous new epoch of the Anthropocene."

— *Clive Hamilton, Professor of Public Ethics, Charles Sturt University, New South Wales, Australia*

"Creatively and persuasively, this book demands an opening-up and a freeing of our modes of thinking and dialogue. In a world of ever-more pressing challenges, this is a powerful anthem for the emergence of the multidimensionality and mutual respect we will all need to live comfortably with each other and with non-human nature."

— *Melissa Leach, anthropologist and Director, Institute of Development Studies, University of Sussex, UK*

"Post-Normal Science is required whenever facts are uncertain, values disputed, stakes high and decisions urgent. It is gratifying to me that the authors have found the ideas of Post-Normal Science illuminating for steering us through this uncertain world. From this book I have gained a deeper insight into what Post-Normal Science can mean for the future."

— *Jerome Ravetz, co-originator of Post-Normal Science, Fellow of the Institute for Science, Innovation and Society, University of Oxford, UK*

"*Independent Thinking in an Uncertain World* alerts us to the need to be creative and open-minded, and to bring the whole of our thinking, the social, ethical, aesthetic and sympathetic, as well as the rational/scientific, to making our decisions about how we live in the World."

— *Wendy Rainbird, International Council of Women*

"Global problems like climate change require global solutions, right? Maybe not. Maybe the answer lies in the resilience that comes from the complexity and inter-relationships in open minded independent thinking by individuals. Read this book and be optimistic about the future."

— *Fred Pearce, contributor to* New Scientist, The Guardian, The Independent *and* Times Higher Education

"Here is a treasurable companion and guide! It emerges from humility in the face of uncertainty, and the mental gymnastics of multiple minds and multidimensionality. Surprisingly easy to read, a feast for the imagination, with gifts from multiple learning cycles, it's for those messy and planetary issues of our times."

— *Pierre Horowitz, Professor, School of Science, Edith Cowan University, Western Australia*

"Amidst the struggle and aspiration for a flourishing future, there is great value to be found in thoughtful, grounded, experiential reminders of how others are navigating. This is a practical, poetic and courageous collection, with an appealing hint of contagion. Readers could easily find themselves engaging in just this kind of independent thinking."

— *Margot W. Parkes, Professor of Health, Ecosystems and Society, University of Northern British Columbia, Canada*

Independent Thinking in an Uncertain World

Any effective response to an uncertain future will require independently thinking individuals working together. Human ideas and actions have led to unprecedented changes in the relationships among humans, and between humans and the Earth. Changes in the air we breathe, the water we drink and the energy we use are evidence of Nature – which has no special interest in sustaining human life – looking out for itself. Even the evolutionary context for humans has altered. Evolutionary pressures from the digital communication revolution have been added to those from natural systems. For humans to meet these challenges requires social re-organization that is neither simple nor easy.

Independent Thinking in an Uncertain World offers workable, field-tested strategies from the frontiers of creating a viable future for humans on Earth. Based on research results from hundreds of social learning workshops with communities worldwide, many of them part of Australian National University's Local Sustainability Project, authors with diverse interests explore the gap between open-minded individual thinking and closed socially defined knowledges. The multiple dimensions of individual, social and biophysical ways of thinking are combined in ways that allow open-minded individuals to learn from one another.

Valerie A. Brown AO, BSc MEd PhD, Director, Local Sustainability Project, Human Ecology Program, Fenner School of Environment and Society, Australian National University and Emeritus Professor of Environmental Health, Western Sydney University has published widely on collective thinking and social change, notably *Tackling Wicked Problems Through the Transdisciplinary Imagination*.

John A. Harris BSc MSc PhD is a past Head of the School of Environmental Studies, University of Canberra, an environmental educator and an action researcher with the Local Sustainability Project and the Alliance for Regenerative Landscape and Social Health, Australia. He is author of *The Change Makers* and co-editor of *Tackling Wicked Problems*.

David Waltner-Toews BSc DVM PhD is Professor Emeritus of Population Medicine at the University of Guelph and has authored or co-authored several texts, more than 100 peer-reviewed papers and a dozen books of poetry, fiction and popular science. He was first president of Veterinarians Without Borders–Canada and a founding member of the Community of Practice for EcoHealth–Canada.

Independent Thinking in an Uncertain World

A Mind of One's Own

EDITED BY
VALERIE A. BROWN, JOHN A. HARRIS AND
DAVID WALTNER-TOEWS

Routledge
Taylor & Francis Group
LONDON AND NEW YORK

earthscan
from Routledge

First published 2019
by Routledge
2 Park Square, Milton Park, Abingdon, Oxon OX14 4RN

and by Routledge
52 Vanderbilt Avenue, New York, NY 10017

Routledge is an imprint of the Taylor & Francis Group, an informa business

British Library Cataloguing-in-Publication Data
A catalogue record for this book is available from the British Library

Library of Congress Cataloging-in-Publication Data
A catalog record has been requested for this book

ISBN: 978-1-138-38721-8 (hbk)
ISBN: 978-1-138-38722-5 (pbk)
ISBN: 978-0-429-42640-7 (ebk)

Typeset in Univers
by codeMantra

Contents

Figures

Tables

Boxes

Contributors

Kathryn E. Andrews BSc PhD. University Fellow, School of Indigenous Futures, Charles Darwin University; Visiting Lecturer, Fenner School of Environment and Society, ANU. Kate chairs Territory Natural Resource Management and is an advisor and consultant. She established the Lake Eyre Basin Coordinating Group and became its first CEO. At Land & Water Australia, Kate established the Knowledge for Regional NRM programme and is a member of the Australian Landcare Council 2010–2015.

Kerry A. Arabena BSocWork PhD. Professor of Indigenous Health, University of Melbourne. Kerry holds the Chair of Indigenous Health at Melbourne University. Her doctorate at ANU won the 2009 Crawford Prize. She has been CEO of the Lowitja Institute and the Chair of Indigenous Health at Monash University. Author of *Becoming Indigenous to the Universe*.

Craig A. Ashhurst BArts GradDipEd CertIVTraining & Assessment, MPhil (by thesis). Director, Niche Thinking Pty Ltd. Craig has been bookseller, youth worker, teacher, developer of educational software and management consultant. Since 2013 he has been a PhD candidate at ANU. Craig formed his own company in 1995 and works in both the private and public sectors.

Valerie A. Brown AO, BSc MEd PhD. Director, Local Sustainability Project, Fenner School of Environment and Society ANU 1995–2015 and Emerita Professor of Environmental Health, University of Western Sydney. Valerie is the author and editor of ten books, including *Tackling Wicked Problems Through the Transdisciplinary Imagination*.

Elizabeth A. Clarke BAgSc GradDipComm PhD (ORCID 0000–0001–5018–810X). Currently Research Fellow Leuphana Universität and Western Sydney University and Honorary Lecturer, ANU. Liz is a systems thinker, transdisciplinary researcher and practitioner of co-design for positive change for sustainability. Her work includes food systems, rural development, environmental management and natural resource management.

Dianne E. Dibley BA LLB DipEd Cert 4 Indigenous Mentoring PhD under examination. Di is an environmental lawyer working on environmental policy and advocacy. She has been National Policy Director Environment & Energy for the Australian Industry Group and National Policy Director for Greening Australian.

Dena Fam BIndustryDesign PhD. Associate Professor and Research Director, Institute for Sustainable Futures, University of Technology, Sydney; Core Member, ISF – Institute for Sustainable Futures. Dena Fam is a system thinker and transdisciplinary collaborator at the Institute for Sustainable Futures at the University of Technology, Sydney.

Rebecca Freeth BSocSc BPhil MPhil Leuphana University Reos Partners. Rebecca Freeth researches, writes about and facilitates collaboration. Always a hybrid herself, Rebecca moves between the worlds of practice, teaching and academia and is based mostly in South Africa.

Mark W. Gardner BAgriSci (Hons) Certified Holistic Management Professional. Educator (Savory Institute). Member AICD; Owner Director, Vanguard Business

Services, Dubbo. Mark is a holistic management consultant and worked as a Savory Institute Holistic Management Educator for 22 years. His Vanguard Business Services has won a number of Regional Business Awards.

John A. Harris BSc (Hons), MSc PhD. Action researcher Local Sustainability Project, ANU; Board of Alliance for Regenerative Landscapes and Social Health. He has published on marine fish, bats and cave ecosystems and urbanization in Papua New Guinea. John established foundational courses in environmental science at the University of Canberra 1972–1999 and postgraduate curriculum. He is co-author and co-editor with Val Brown of three books. His own book is *The Change Makers*.

Dana Kelly BSc MSc PhD. Dana is a social scientist undertaking research in the Australian rangelands and outback. Her consultancy is in community engagement, social change, organizational collaboration, agricultural extension, protected area management, monitoring and evaluation. She is an Associate of the International Water Centre, University of Queensland.

Judith A. Lambert BPharm BSc(Hons) PhD, GradDipEnvManag, GradDipBusinessAdmin. Judy's training and career span environmental, social and medical sciences and business administration. With the late Dr Jane Elix, she formed a consultancy business, Community Solutions, some 25 years ago. Another collaborator is Professor Valerie A. Brown.

David S. Marsh MSustainableAgriculture. Regenerative farmer, Allendale, Boorowa, NSW. National Landcarer of the Year 2018. An award-winning regenerative farmer, professional speaker and author; Founding President of the Boorowa Community Landcare Group, 1989, committee member for 15 years. Since 2014 he has been Coordinator of and published blogs on the Alliance for Regenerative Landscapes and Social Health website (ARLASH.com).

Charles J. Massy OAM BSc, PhD. Farmer, Severn Park, Cooma, NSW, Australia. Charlie graduated in Zoology in 1976, before developing a Merino sheep stud (Severn Park). He manages a grazing property in NSW while consulting in Merino breeding and landscape design, and regenerative agriculture. Author of *Call of the Reed Warbler: A New Agriculture – A New Earth*.

Richard M. Thackway BSc MSc. Adjunct Fellow, Fenner School of Environment and Society, ANU and School of Physical, Environmental and Mathematical Sciences, University of New South Wales at the Australian Defence Force Academy; and research scientist with extensive experience in science policy agencies in the Australian government.

Ellen Wall PhD. Ellen is retired from environmental sciences at the University of Guelph. A rural sociologist, she has managed a national research network on climate change impacts and adaptation for agriculture (C-CIARN-Agriculture). She was an expert reviewer for the North American chapter of the IPCC's Fourth Assessment Report.

David Waltner-Toews BA DVM PhD. A veterinarian and epidemiologist specializing in zoonotic diseases and ecosystem approaches to health; Professor Emeritus at University of Guelph and adjunct researcher at Centre de recherche interdisciplinaire sur le bien-être, la santé, la société et l'environnement (CINBIOSE), University of Quebec at Montreal. Author and co-author of several textbooks, over 100 peer-reviewed papers and book chapters.

Julia H. Wolfson MAppSc DipPW PhD. Accredited – International Association of Process-Oriented Psychology. Principal, Turning Forward and Associates; Founder and

Principal, International Deep Democracy Institute. Faculty Eldership Academy; Visitor, Rudolf Steiner University College, Oslo. Her recent book is *Applying Deep Democracy in Human Services.*

Acknowledgements

The three editors wish to acknowledge the following contributors to *Independent Thinking in an Uncertain World: A Mind of One's Own*.

First, we thank the members of the Fenner Writing Group, Fenner School of Environment and Society, the Australian National University. As fellow members of the writing group, we (Val and John) thank you for the four years we all spent getting to know one another and generating our own examples of independent thinking. Members of the writing group, Kate Andrews, Kerry Arabena, Craig Ashhurst, Liz Clarke, Di Dibley, David Marsh, Charlie Massy, Richard Thackway and Julia Wolfson, provided the chapters for Part II of this book. Thanks also to Professor Stephen Dovers, former Director of the Fenner School who first agreed to host the writing group.

Thank you to Dana Kelly, Gil King, David Marsh and Wendy Rainbird, the four readers who shared their ideas in valuable discussions as well as in a more formal review. From their diverse experiences, they made excellent suggestions that we have universally adopted. The final product is, of course, our responsibility.

Many individuals have contributed their skills to the production of the book. Our editor and friend, Paul Wallace, worked wonders on our text, way beyond the call of duty. Craig Ashhurst rescued the illustrations. The co-authors, Mark Gardner, Dena Fam and Rebecca Freeth, supported Writing Group members. Routledge editors Rebecca Brennan and Leila Walker were supportive and gave us sound advice. The National Botanic Gardens provided a sympathetic background for our writing group discussions.

Val writes: Judy Lambert and I were fellow facilitators at the workshops that provided much of the material for the book. To the many hundreds of members of the Local Sustainability Workshops from all over the world, 1995–2015, thank you for your willing participation and your commitment to whole-of-community change. I also wish to acknowledge the Human Ecology Forum of the Fenner School, which provided a haven for sharing ideas among independent thinkers over many decades.

To my fellow editors, thank you for the warmth and strength of our collaboration. To my husband Robin, your memory and influence continues. My children, Matthew, Bina and AJ, provided love, support and encouragement while I concentrated on this book. Hopes that my four grandchildren, Chris, Sarah, Elliot and Amon, would live in a better world drove me on.

John writes: The companionship, advice and patience of my wife Jane has been unwavering and vital to me and I thank Craig, Stephen and our granddaughters, Brooke, Ellie, Grace and Scarlett who help to make it all worthwhile.

I wish to acknowledge the University of Canberra and the many students in environmental science I taught there for their enthusiastic questioning stance that kept me learning and open minded. The Human Ecology Forum at the Australian National University was also an important stimulus to me for open discussion on diverse research topics over many decades.

I wish to thank my parents who supported my education and set me free to discover the creek mentioned in the Prologue. My thanks also to the late Allan Fox, naturalist and author, who never tired of sharing his stories and encyclopaedic knowledge on our many trips outback inspired by the beauty of the Australian continent.

I am indebted to Val for inviting me to share her search for a better world. It has been a challenging, joyful and demanding partnership but most of all there was friendship.

David writes: For the many hours we spent working and debating and finding new ways to think about, and act in, the world, to leave it better than we found it, I thank the colleagues and members of the Agroecosystem Health Project at Guelph; the Post-normal Science network (especially Jerry Ravetz and Silvio Funtowicz); the Dirk Gently Gang (including the late James Kay in Canada and Thomas Gitau in Kenya, both of whom died far too young).

I thank the diverse and inspiring members of the Community of Practice for EcoHealth–Canada (Copeh-Canada), including especially the late Bruce Hunter, who helped to keep me grounded; Kerry Arabena and Pierre Horwitz and Margot Parkes for welcoming me as an honorary member of the Oceania Copeh chapter; my many colleagues in the Southeast Asia and East Asia EcoHealth and emerging diseases networks. Thanks to the programme directors and scholars at Canada's International Development Research Centre, for supporting and challenging me.

Thanks to my children, Matthew and Rebecca, and their partners, Steve and Lydia, and their children (my grandchildren) Nikolas, Annabel, Wendell and Ira, who continue to inspire me and teach me new things and give me hope. And Kathy, for your support and forbearance of my international and intellectual obsessions and the many times you kept the home and household going and the kids fed when I was away in Kenya, Colombia, Nepal or Thailand.

The epigraph by Michael Ignatieff at the beginning of Chapter 4 from *The Needs of Strangers* (New York: Penguin Books) is ©Michael Ignatieff, 1984, and its reuse here is by permission of the author.

The epigraph by Larry Haworth at the beginning of Chapter 4 from *The Good City* (Bloomington: Indiana University) is ©Indiana University Press, 1963, and its reuse here is by permission of the publisher.

Foreword

Jerome R. Ravetz

I feel that it is an honour and a privilege to contribute a foreword to this inspiring volume. It is very gratifying to me that the authors have found the ideas of post-normal science to be illuminating. And from them I have gained a deeper insight into what post-normal science can mean.

First, of course, this volume records the multiple achievements of Valerie Brown in creating new conceptions of the way we think and act, individually and collectively. She has not only achieved new insights and described them; equally importantly, she has embodied them in practical methods for their further development and use. Of course, any simplified scheme can be abused in a box-ticking culture; but when such a scheme expresses deep truths about a system, it can enrich and even transform practice. Her list of the dimensions of mind, deployed in conjunction with her list of varieties of actors, provides participants a means of achieving self-awareness. Thereby, as she and her co-editor John Harris know so well, the most intractable obstacle to creative common endeavour, the inability to see oneself from outside, can be overcome. In that way, the most important task facing our civilization, the employment of nonviolence in policy, can be achieved. This, I believe, is the great achievement of this volume.

Articulating an enriched, complex vision of oneself in society is challenging. Contradictions and hurts are everywhere. We can carry on better when we see how others have overcome them, both in group work and internally. My friend, comrade and mentor David Waltner-Toews is a living example of that struggle. His histories of successful campaigns, together with personal reflections and poetry, all integrated in his personal vision, show us that it can indeed be done; and that integrity and hope can indeed be maintained through the inevitable messes of actions to make the world a better place.

Kerry Arabena provided a precious occasion for me to reflect on myself and my work. It started with her focus on the vision of nature of indigenous peoples; this could be described as including all the dimensions of mind in Valerie Brown's scheme, where the aesthetic and sympathetic are as essential as the physical and where historic crimes are part of the present and future. From this perspective, the poverty of 'normal science' suddenly becomes clear. We could think of it as the social perception of a world where everyone is colour blind, tone deaf and afflicted by anosmia and ageusia. In policy terms, one can think of words used in South Africa during the debate on ending apartheid: those willing to change were *verligte*, enlightened, while the opponents were *verkrampte*, cramped. At the extreme, there is the old joke about science having disproved the existence of consciousness, with the obvious exception of white male Harvard psychology PhDs. This is the implicit indoctrination of students in all officially certified sciences. It prepares them to believe that they can manage and change the world in that benighted image.

Now I better understand the significance of post-normal science. Less important are the mantras describing the condition. The essential message is: yes, it can be different.

Science can encompass diversity, variety, love and hate and tragedy. Doing real-world science can and must involve conflict and reconciliation and knowledge of self and others. Kerry's term 'post-normal reconciliation' expresses the spirit of post-normal science as I have now come to understand it; it is to be distinguished from the 'normal' sort whereby those oppressed Others have a chance to become real imitation white folks.

I need not go on at length at the other contributions to this volume, particularly those about the practical experience of the farmers. One might say that a key feature of modern societies is the way that they provide a shield and buffer from the workings of the external world. For the last half-century, ever since *Silent Spring*, it has been common knowledge that our science cannot totally dominate Nature. And on an increasing variety of fronts, our technological successes are realized to be producing their own negations, as with superbugs and superweeds. On the American Great Plains, the Ogallala aquifer is being pumped dry, with a future lifetime estimated as a few decades. Perhaps the name will last longer than the expropriation? It could be that in Australia, where the most sophisticated governance and technology confronts a very alien ecosystem, vital lessons about conservation and regeneration will have a better chance of being heeded. Our authors should be recognized eventually as heroes of survival; and perhaps they will.

Finally, I want to recognize the other achievement, the one that made all this collective work possible: the Fenner Writing Group. Although this was not about science, it was clear that same inspiration, which I am now honoured to call 'post-normal', made it such a success.

Thanks to the authors and particularly, of course, to Valerie Brown, for their achievement and for giving me the opportunity to share in it and learn from it in this way.

Prologue

The bat cave

John A. Harris

This is my story of living through the very changes in thinking that are at the heart of this book. Growing up in the foothills of Adelaide, South Australia, with a permanent creek flowing past my home, it seems inevitable that I would fall in love with natural things. And, once my curiosity was sparked, I wanted to understand more about how my tiny world fitted into the larger world. To fulfil my dream I went to Adelaide University and became a practising scientist. I began to think for myself. I had my first inkling of what this meant during my PhD research in a bat cave.

Bats are true flying mammals. In my case, they were like tiny flying mice. They lived in the bat cave by day and left to feed on insects at night. By their presence, they altered the cave temperature and created an environment of their own. I studied a small inner chamber where I entered the chamber through a low crawl space by torchlight. If bats were roosting there, they brushed past me as they left. Understanding the interrelationships between the bats and the cave was the key to unlocking the secrets of the cave. It was also the key to understanding the human world and myself.

The bat cave had seemed to be a simple system in a stable temperature 'laboratory'. However, this idea had more to do with my scientific thinking. I had misunderstood the interrelationships in the cave, as I had failed to understand my own social system. It was only after I began to think for myself that my mind was set free and open to learn from these mistakes.

Drawing on my own thinking allowed me to see how the bat cave exemplifies the extraordinary diversity generated by the evolutionary process of life on Earth; a process that included my own thinking. The bat cave demonstrated the ever-changing character of the Earth. This was my first inkling of a new order in my mind. It took me more time to know intuitively that I was a part of that interdependent world. I was working late one night when, suddenly, I felt connected to the whole of life on Earth. It brought a powerful emotional response.

These flashes of insight came to me when I was grappling with a tension created by the conflict between my commitment to objective science and my mind-changing experience in the cave. I abandoned the goal of a quantifiable understanding. Instead, I worked tirelessly to sort out the confusion in my mind. It was that sustained focus and attention that helped free my mind, allowing all dimensions of my thinking to fire and flash. I got much better at working creatively, using my intuition and imagination as well as my scientific training, and to use them all together.

When I set out on my PhD, I was a research scientist focused on objective thinking about the physical and biological conditions in the bat cave. To gain access to the cave, I needed the help of my social colleagues, local cave enthusiasts and landholders. My love for natural systems led me to a new ethical position: to reduce the extent of my disturbance to the bats and to increase my responsibility to my society. This led me to developing sympathy for the tiny bats and so for other vulnerable beings. I realized how much I had become a part of the system I was studying. I had experienced my mind being able to use its physical, social, ethical, sympathetic and aesthetic dimensions and remaining whole at the same time.

The bat cave changed me forever. I found myself in a new story about my place in the natural and social worlds and the interdependence within and between these worlds. For years, I thought that was the story of the bat cave. It took me a long time to realize that was my story, too. The bat cave had allowed me entry to multidimensional thinking. I see my tiny world as multifaceted and whole at the same time. This is an enduring story and it runs through this book.

It is a story of the world I fit into, for better or worse. It is not only complex and beautiful; it generates a vast natural and cultural variety. There is a capacity for renewing the panoply of planetary life in unexpected and sustaining ways that are beyond my powers of understanding. Nevertheless, my journey to understand has brought me to the strength of the multiple dimensions of an individual mind and this book.

Part I

Ideas

1

Thinking for oneself

Outside the square

Valerie A. Brown

Every now and then a new idea spreads throughout human society. The collected thinking of the world slips a cog while it absorbs the new idea into its everyday life. We humans have already faced the dramatic changes of the agricultural and industrial revolutions. Both of these permanently altered the relationships that link individual humans, their societies and the Earth. A revolution of our own times brings with it the challenge to establish new long-term collaborations between humans and humans and between humans and their only planet. For this, we look to a mind fit for the task.

Just this morning, my social media carried pictures of a warming planet: a free-floating iceberg as big as a small city, ice sheets crisscrossed with large cracks and a starving polar bear. The President of the United States, the most powerful man in the world, tweets that the world is getting cooler. Facebook's founder moans that his wonderful tool is being used to service a confused world of 'alternative facts'. Floods and droughts predicted for over 30 years still catch people unprepared. Refugees from intolerable living conditions are shunted from place to place, with no foreseeable solution. Governments fail to face up to the extent of the social and environmental change. I, as an individual, am left to interpret this complex world for myself.

There is no shortage of confident explanations of this complexity. Professions, disciplines, political parties and people in the street all have their own versions. More important is to ask, what is the question? Nobody seems to be sure. There is, however, one question that I keep asking myself. In these uncertain times, why aren't we drawing on the immense capability of our own minds? Some people are singled out to give us the answer, some as geniuses, some as powerbrokers, others as heroes. The rest of us are expected to leave the too-hard questions to them. As human beings, however, they are no different from any of us. We all have the same brain structure and ability to think that allows us to think for ourselves. To go back to a question about the question: How can we draw on the full capacity of our minds in addressing the issues of our time?

Climatologist Will Steffen has argued that the formal title for our time in history should be the Anthropocene. The 'Anthro' part is a reminder that human ideas have led to unprecedented changes in our social lives and physical environment. Changes to the air I breathe, the water I drink and the energy I generate are our environment's response to human actions. These changes are not primarily addressed to us humans: They are the Earth looking after itself. As an individual, I must respond on my own behalf, hoping to find companions to help me. However, to address the necessary social reorganization will be neither simple nor easy. I cannot do it by myself: It is too vast a canvas.

Once I applied my mind to the emergent issues, I could see for myself the crucial need to negotiate our way to the solutions, solutions that we can all share with one another and with the world. Otherwise we will all continue to be deaf to this changed world speaking in its own voice. The Earth will activate its own feedback systems whether we like it or not. We need to learn to listen. Communities do the best they can with the resources they have. Worldwide experience of working with changing communities has taught me that there will always be those who try to avoid the issues of the Anthropocene. Others will take a one-eyed view, fixing their minds on one problem and one solution. Belief in one right answer can lead to a single source of authoritarian control. At one extreme, this singlemindedness leads to the mass Jonestown suicides and the terrors of the Nazi regime. At the other, it produces individuals alienated from their own society. The planet has been dealing with immense changes over many billion years. Whether we humans act wisely on the issues or not, the planet will survive. Unless we can escape our society's current focus on one right answer, we humans may not.

After working with communities all over the world, I came to realize that existing organizational structures and current ways of thinking could not be expected to break new ground. They have established interests to protect. This often means that they work actively to prevent change. Why should I be surprised? Key social institutions such as education, law, health and government have been building walls around themselves ever since the 17th-century scientific revolution. Thinking has shrunk to considering only one's own section of the social network. As I have described in previous books, in Western culture our knowledge making has been divided into different specialized compartments, each of which protects itself from change.

The communities I was working with were looking for whole-of-community change. In the next chapter, Chapter 2, I give an account of the research team, design and overall outcomes of a series of collective learning workshops. Here, I describe the steps toward learning to recognize multiple dimensions in each individual's thinking. I was leader of the Local Sustainability Project (LSP) at the Australian National University over 1990–2015, responding to issues of whole-of-community change. Our research teams found that the Local Sustainability Project workshops were successful in bringing diverse groups together on shared tasks. However, when the workshop members went back to their usual social environment, it was a different matter. The old barriers returned. It made me think of the proverb: You can take a horse to water, but you can't make it drink.

During the early workshops the team identified different knowledge cultures developed by individuals, communities, specialists, organizations and holistic thinkers. Each knowledge culture had its own aim, language, content and set of practitioners. To my surprise, the crucible for significant change did not arise from any one of the knowledge compartments. Rather, change came from certain individuals who stood out in each of the workshop groups. These individuals acted independently of their colleagues and thought more broadly across the issues.

Following this unexpected behaviour, the research team and I redesigned the workshops. The aim of the new workshops was to learn more about the individual's desire for independence and their capacity to think beyond the square. In debriefing sessions after a workshop, members of the research team and of the groups identified different dimensions of the individuals' thinking: the physical, social, ethical, aesthetic and sympathetic dimensions of their individual minds. The *physical* dimension was accessed through objective observations and measurements; the *social* dimension by shared stories and symbols; the *ethical* dimension by action on moral principles; the *aesthetic* dimensions by recognition of patterns and style; and *sympathetic* dimensions in the warmth of one-on-one communications.

The recognition of multidimensional thinking demanded further investigation. Once recognized, I began to find signs of the multiple dimensions of mind everywhere. I was in the position of someone who buys a red car and then suddenly sees red cars everywhere, when they had never noticed them before. Once the multiple dimensions of mind had been identified in the workshops, I found these multiple dimensions whenever independent individuals were thinking for themselves. In Chapter 2, I explore the relationships between the closed minds of the knowledge compartments and the open minds of independent thinkers. For now, I report on some examples of the value of the contributions made by multidimensional minds to difficult times (see Box 1.1).

BOX 1.1 A REGION IN CRISIS

A remote region in Australia's far north was suffering a severe drought. Drought is usually associated with a lack of rain, a meteorological drought, a physical measurable dimension. But social life is hard to carry on when the view out the window is red dust and empty paddocks. Politicians and city people focus on the images of dying cattle and sheep, failing to understand that everyone in these remote rural communities is affected. Yet the sense of place is strong as local people love the aesthetics of the red sand hills and open spaces of the region where they live. They are also strongly connected, sympathetic to one another's problems and with a strong ethical tradition for helping one another.

The impacts of drought on the whole region has been devastating. Everyone in this remote community was affected. Farmers and their children,

(Continued)

shearers, fencing contractors, hairdressers and local grocery stores suffered an economic drought, as no money comes into the town. A social drought developed as large numbers of people leave town looking for work. Services decline, schools close; sports clubs struggle to find enough people for a team.

Nevertheless, people changed the way they did things and diversified their businesses. Local people volunteered to keep social events happening. Local charity groups provide credit cards for people to spend locally. Innovative local businesses developed in response to drought, often based on people's love of the area. One cattle family turned tourism operators take visitors on a Cobb and Co coach ride, providing a genuine heritage experience. Strong social capital and networks helped. Individuals wore many hats, volunteering with Rotary, the Drought Committee, helping the homeless, working in schools and sitting on the hospital board.

Comment: Paradoxically, the needs of remote regions are often ignored; yet distance gives them freedom to consider their own and other's perspectives and to think and act outside the square. Independent thinking is especially valued in times of hardship, such as this drought in Longreach, where people were able to access all dimensions of their minds.

Source: Kelly, 2018

The many failures of orthodox methods to resolve the issues of the Anthropocene: carbon emissions, violence in the cities and the obesity epidemic, had already made me wonder whether there was another way. From the LSP workshops, I collected a list of possible ways of responding to the complex issues (see Table 1.1).

Table 1.1 Responses to the issues of the Anthropocene

Responses to the issues	*Implications of the response*
There is nothing we can do	We are but playthings of fate
We did it, we should fix it	We are responsible beings
It is a multidimensional issue	We have multiple ways of thinking
Combine all the responses	We value holistic thinking
Go back to nature	We are part of nature
Planetary resources are infinite	We can take whatever we want
Technology has the answers	We are a rational, logical species

During the Local Sustainability Project workshops, each of the options in Table 1.1 had its own champion. There were hot debates within each option: right versus left politics; community sociology versus classical economics; industrial versus regenerative farming; specialist versus holistic thinkers. Different responses to social and environmental change brought fiery advocates into the fray. No wonder people rolled their eyes in frustration. Some workshop members expected that those in power would accept the responsibility of making the necessary decisions. Others looked for fresh ideas from separate knowledge cultures, only to find that each knowledge culture rejected the solutions suggested by the others. Now realizing the multidimensional minds each individual brought to the table, we asked workshop members to introduce the idea of 'and' in the list in Table 1.1 and so bring their energy to understanding each other: collaboration rather than confrontation.

When the project team asked the independent workshop members to switch from the idea of 'or' to the idea of 'and' to the list in Table 1.1, this produced an unexpected response: combining the options. Workshop members agreed that each option had some value and held some element of truth. The idea of combining all the options, of introducing 'and' between them was considered strange, although not unreasonable. Once the 'and' was introduced, general discussion revealed the same multiple dimensions of thinking in individuals in previous workshops. The responses in Table 1.1 included that resources are endless (physical dimension); we can take what we choose (social dimension); we are responsible beings (ethical); we appreciate universal patterns (aesthetic); and we are part of nature (a sympathetic/unsympathetic world). Once combined, the responses stood out as multidimensional and collaborative, making good use of the 'and'.

As the multidimensional learning from the workshops grew, I tried to apply my own multidimensional mind to major social changes and their individual participants. My list included the international space race (with highly trained humans walking on the moon); reunification of a divided nation (both sides dancing on the Berlin wall); collaboration among normally competing multinational organizations (humans protected from the polar ozone holes); and the birth of the open source worldwide web (explosion of the use of Facebook and Twitter). A closer examination of these events uncovers a consistent pattern in their success. Each of these events was the result of individuals acting independently outside the bounds of their society. Each of these changes involved individuals drawing on the multiple dimensions of their minds.

Physical, social, ethical, aesthetic and sympathetic ways of thinking made up the individual mind. Going back to Table 1.1, each of the options for responding to issues of the Anthropocene, considered separately, could lead to conflict and denial. The same problem arises in considering the multiple dimensions of mind. The common practice is to expect each individual to draw preferentially on only one or two of the dimensions. Individuals are asked to think objectively, socially, ethically, aesthetically or sympathetically, according to the context. This is where 'and' is needed again. Human minds are whole, not divided. As we shall see in the examples that follow, when all the dimensions are involved, a synergy develops that embraces the whole of an issue. There are, however, serious barriers in the way of bringing them together.

Current Western society separates the physical, social, ethical, aesthetic and sympathetic dimensions of mind. Although all dimensions of mind are present in everyone, everyone's thinking is asked to be primarily objective. Since the scientific revolution in the 17th century, there has been pressure to interpret the world through impartial observations. The revolution was called the Enlightenment from the glaring light of its objectivity being directed onto one thing at a time.

The result was summed up by William Wordsworth in 1798:

> The primrose by a river's brim
> A yellow primrose was to him
> And it was nothing more. (*Peter Bell: A tale*)

This small yellow flower is variously considered as a physical object, the first sign of spring, a thing of beauty, a unit in an ecological niche, a botanical title, a symbol of England, a step in the evolutionary story, a source of inspiration, a fellow form of life, a metaphor for freshness and innocence and more. Our minds are capable of holding all these ideas in our head at the same time. Yet the rich fund of ideas is usually reduced to one or two from the long list of possibilities.

This culture of objectivity, the reliance on physical evidence alone, has led us to ignore many of the ways we can think about a primrose or anything else for that matter. As a result of the Enlightenment, logical thinking and physical observations have been preferred to creative thinking and sensitive understanding for over 400 years. The monopoly of objective thinking devalues not only the primrose but much of our lives and our ways of thinking. At the same time, 17th-century science began an era of brilliant discoveries, from the mechanical inventions of the industrial era to the spaceflights and nuclear power of the Anthropocene. With this track record, it is no wonder that science alone has been expected to solve the world's problems.

In the early Medieval times, about 700 AD, about 1000 years before mathematician Rene Descartes launched the principle of the objective mind, a Franciscan monk, Saint Boniface, wrote of the dire results of separating 'the eye of the mind from the eye of the heart and the eye of the spirit'. The state of the Anthropocene in the 21st century suggests that they were both right. There is need to accept and adapt the heritage of the Enlightenment for individuals to understand themselves and the issues that they face in the Anthropocene.

In 1972, two experienced social planners, Horst Rittel and Melvin Webber, went beyond solely objective approaches in examining the complex problems of their era. They labelled such problems 'wicked', not because they were morally wicked. Their wickedness rested on the level of difficulty in bringing change to the society that generated the issues in the first place. Wicked problems span the earth and individuals have become lost in the tailwind. The idea of wicked problems spread quickly across fields such as engineering, medicine, education and research. Wicked problems have the following characteristics. Their resolution:

- requires changes to the society that caused them
- involves multiple interests and so has multiple answers

- calls for collaboration, so it can have no one right answer and no single answer is right or wrong, good or bad
- brings changes to a society so that there can be no final answer, only the best that can be done at the time
- eventually rests on individuals and so needs to include the multiple dimensions of their minds.

The originators of the idea of a wicked problem added a postscript: While a wicked problem may not be morally wicked, failure to take account of complexity and its social implications *is* morally wicked. For individuals to fulfil this challenge to address the full complexity of an issue, they will need to draw on the whole of their mind. To solve a wicked problem one needs to determine the physical context of an issue. Since resolution of the issue requires changes to the society itself, individuals will become critics of the established social order. With multiple interests, there cannot be one right answer and so ethical principles come into play and, with no final solution, individuals, like artists, will need to draw on their creativity and imagination. Individuals taking on the pressures of working with change can hope to find sympathy and friendship among their colleagues. As an example of the multiple dimensions of a wicked problem I have collated a table on the global obesity epidemic (Table 1.2).

My experiences in countries as different as Australia and Nepal leads me to the conclusion that if individuals are to contribute to resolving the wicked problems of this uncertain world, they will need to look at the issues afresh. This is not a simple matter. In any country, changes in practice will challenge the ruling power structure in which experts and organizations vie for the top rank and individuals are relegated to the bottom. We are likely to accept our social labels without question, forgetting, or never even knowing, that they are not absolute. The tight

Table 1.2 Multiple dimensions of the obesity epidemic

Physical	Increased consumption of high energy sugars, little exercise, high level of craving, individual metabolism
Social	Incidence greater in low income populations; fewer physical jobs; instant gratification, strong marketing
Ethical	Unethical marketing of sugar products to children, responsibility to self and others
Aesthetic	A current preference for thin individuals, nineteenth century preference for heavy ones
Sympathetic	Shared mutual understanding of the difficulty of dieting, recognition of the same risks, identity with others of one's own shape

allocation of roles seems to me as sad as it is unnecessary. If we do not reflect on our own place in society, we will not be aware of our own assumptions that limit our thinking. Although individuals may accept the cultural allocation of polarized selves, their individual experience tells them that those roles are not set in concrete. No matter how rigid the allocation of social roles may be, my own experience tells me that, while I may appear to live my life in ways that have been socially assigned, those roles are not my real self.

I find examples of individuals acting independently everywhere I look, even in the myths of ancient Greece. The 12 labours of Hercules have long been symbolic of an individual's triumph through their physical strength. On the contrary, his successes were a matter of using all of his mind. Cleaning the filthy Augean stables, he did not dig a ditch – he diverted a river (physical dimension). Sent to kill a lion terrorizing the countryside, he cornered and strangled the lion, saving the villagers (social dimension). Expected to fight the warlike Amazons, he went in peace (ethical dimension). Bringing back the beautiful golden hind, everyone was in awe (aesthetic dimension). The grateful villagers asked the gods to adopt him and so take him away from danger (a sympathetic act). Hercules was using all dimensions of his mind; different dimensions stand out in different contexts.

Contexts as far apart as a Greek myth and a drought-stricken region in Australia confirm the value of drawing on all multiple dimensions of mind (Box 1.1). Stories of Lincoln, Gandhi and Mother Teresa hold parallels to that of Hercules and his so-called impossible tasks. Each of these people has become a symbol for using the whole of their mind to achieve the apparently impossible task of bringing change to an entire nation. They also stand as symbols of an individual's triumph over the controls established by society. So far, my reflection on the multiple dimensions of mind has been projected outwards on resolving problems. It is also important that I turn my reflections inward. Thinking about one's own existence (that is, introspection) brings with it some valuable gifts. Although I may accept the cultural allocation of a polarized self, my own experience tells me that my roles are not set in concrete. I can choose for myself.

In my own experience, inward reflection brings with it many gifts. This allows me to know myself better. It identifies the assumptions that will affect any decision I may make. Inward reflection can help open up previously unrecognized dimensions of my own thinking. Most valuable of all, inward reflection resolves the edict of Socrates: 'An unexamined life is not worth living.' The tension between socially constructed ways of thinking and individuals thinking for themselves is as old as records of human existence. Turning the reflections inward, a thinker can recognize the dimensions of their own thinking. For myself, introspection has allowed me to know myself better; identify assumptions that I did not know I had; and gave me access to all the dimensions of my thinking.

However, whenever I reflect on my own thinking, I am breaking many social taboos. Time spent on introspection can be dismissed as egotistical, neurotic, selfish and self-centred. On the other hand, it might help lead into the new ways of thinking needed for the Anthropocene. In Table 1.3, exploring all dimensions means that my physical self will have been identified through my physical characteristics. My social identity has been determined by my society, derived from my shape, gender, personality, learning and my cohorts' idea of wealth. My position on ethical matters, telling right from wrong, will have been established during childhood and amended as I learnt from life events. My sense of the beautiful seems to be built in, influenced by my society and experience. Finally, feelings of sympathy arise when I feel deeply for somebody else (see Table 1.3).

Orthodox psychologists and philosophers have given considerable attention to individual and independent

Table 1.3 Recognizing one's multidimensional self

Physical dimension	Objective measures of my body shape, height, weight, health status, sex, life expectancy …
Social dimension	Relationship to my society's ideas of 'normal' shape, gender, aging, personality, income, beliefs, politics …
Ethical dimension	Moral principles underpinning my actions: humanism, Buddhism, Moslem, Christian, other authorities …
Aesthetic dimension	My sense of 'fit': beauty, ugliness, proportion, design …
Sympathetic dimension	My feeling for, sharing with, supporting, loving someone else …

thinking. Their contributions are explored further in Chapter 3. I discuss a few of the most relevant here. Howard Gardner, a developmental psychologist working in education, identified forms of intelligence that fall outside IQ tests. There is considerable overlap between Gardener's multiple intelligences and the multiple dimensions of thinking we have identified in this chapter (see Table 1.4). However, intelligence, defined as the use of the rational mind, is only one part of each dimension of our minds. Each dimension also includes creativity, intuition and logic.

Educational psychologist David Kolb's interest was in the relationship between ways of thinking and the learner. Kolb and later researchers linked ways of thinking with the thinker's role in society.

Scientists were likely to emphasize physical thinking, administrators socioeconomic thinking and community workers sympathetic thinking. A 1995–2015 project of the Australian National University, the Local Sustainability Project, made use of Kolb's experiential learning circle to identify the social constraints that limited the thinking of individuals who wanted to think for themselves.

Philosophers of ideas have been interested in ways of thinking for centuries. A Western account of thinkers and their thinking must, of course, begin with Plato. Plato is credited with drawing on physics, ethics, politics and the arts to explore how human ways of thinking create knowledge.

Table 1.4 Dimensions of mind and related intelligences

Dimensions (Brown)	*Intelligences (Gardner)*
Physical	Naturalistic
Social	Interpersonal
Ethical	Inter- and intrapersonal
Aesthetic	Rhythmic, visual
Sympathetic	Intrapersonal

He concluded that the most that humans could know was a pale imitation of an eternal and unchanging world of forms. Later philosophers explored the multiple ways of thinking that grounded that pale imitation of life.

Sociologist Helga Nowotny considers that there was a need to bring all ways of thinking together to create 'socially robust knowledge'. She suggests that robust knowledge would then be strong enough to introduce human concerns into what had been limited by planners and politicians to a physical domain. The broadening of their thinking in creating fresh knowledge would then directly relate to how people wanted to live. Philosopher of ideas, Isiah Berlin, suggests that ethics involve the pursuit of the ideal. He describes ideals as the set of principles by which people direct their lives.

In pursuit of an aesthetic way of thinking, in his "Ode on a Grecian Urn," the poet Keats writes: 'Beauty is truth, truth beauty, that is all ye know on earth, and all ye need to know.' His writing on the aesthetic imperative in all thinking is a powerful one. Martin Buber brings the subtle emotional relationships between two people to our consciousness in *I and Thou*. In all European languages, except for English, the words for 'I' and 'thou' distinguish between a formal and an intimate relationship. These examples of philosophical inquiry are variously called transdisciplinary (respecting all ways of thinking in their own right) integrative (combining ways of thinking in a single form of knowledge) and common sense (each individual's local experience).

Earth's responses to human actions are not trivial. Changes to energy sources, ocean temperatures, chronic diseases and food supplies are changing the ways human beings are living, all over the planet. Human societies have not had to face such dramatic changes in their short evolutionary history. The next task for humans of the Anthropocene is to respond to the challenges through changes to the physical state, social structures, ethical principles, aesthetic designs and sympathetic understanding of the world we all live in. If it is possible to achieve collaboration between independent individuals who reflect on all these dimensions in their own mind, there is a chance. The Anthropocene may be an epoch where we all share in creating a self-healing environment and a people-friendly, just, wondrous and caring world.

In Part I, ideas that inform the process of multidimensional thinking are explored. Chapter 1 contains stories of individuals who applied the multiple dimensions of their mind to issues their society couldn't resolve. One of the consequences of independent thinking is the need for independent voices. Each of the authors in this book has been free to use his or her chosen medium: text, poetry, diagrams, maps, boxes, tables and figures. You will notice that in my own chapters, I chose not to use direct referencing, offering instead a general bibliography in line with my use of the first person, the 'I'. In the more formal chapters, there is full academic referencing.

In Chapter 2, a review of collective learning workshops describes how the workshops generated mutual learning of the fixed social and the open independent ways of thinking. Chapter 3 considers independent individual minds in some depth, as well as avenues for bringing them together as an undivided mind. In Chapter 4, the theme is the importance of the relationships between independent individual thinking and public well-being. Chapter 5 contains an account of the experiences of a writing group, whose members demonstrate the principles of collective multidimensional thinking in action. Chapter 6 foretells the ways in which Chapters 7–16 that make up Part II offer frameworks, modes of inquiry, a language and a story of multidimensional thinking for an uncertain world.

Bibliography

Berlin, I. (1998) *The Proper Study of Mankind: An Anthology of Essays*. London: Pimlico.

Brown, V. A. (2008) *Leonardo's Vision: A Guide for Collective Thinking and Action*. Rotterdam: Sense.

Brown, V. A., Harris, J. A. and Russell J. Y. (2010) (eds.) *Tackling Wicked Problems: Through the Transdisciplinary Imagination*. London: Earthscan.

Brown, V. A. and Lambert, J. A. (2013) *Collective Learning for Transformational Change: A Guide to Collaborative Action*. London and New York: Routledge.

Brown, V. A. and Harris, J. A. (2014) *The Human Capacity for Transformational Change: Harnessing the Collective Mind*. Abingdon: Routledge.

Buber, M. (1958/2000) *I and Thou*. Ronald Gregor Smith (tr.), 2nd edn. New York: Scribner Classics.

Gardner, H. (1983) *Frames of Mind: The Theory of Multiple Intelligences*. New York: Basic Books.

Gardner, H. (1993) *Multiple Intelligences: The Theory and Practice*. New York: Basic Books.

Hamilton, E. (1962) *Mythology: Timeless Tales of Gods and Heroes*. New York: Mentor Books.

Keats, J. (2010) *Poems of John Keats: Popular Penguins*. London: Penguin.

Kelly, D. M. (2018) *Beyond the DUST: Impact of Drought on Town Businesses in Central West Queensland and some solutions*. Longreach, Report for the Western Queensland Drought Committee.

Kolb, D. A. (1984) *Experiential Learning: Experience as a Source of Learning and Development*. Englewood Cliffs, NJ: Prentice-Hall.

Kolb, D. A., Lublin, S. and Spoth, J. (1986) 'Strategic Management Development: Using Experiential Learning Theory to Assess and Develop Managerial Competencies', *Journal of Management Development*, 5(3), pp. 13–24.

Nowotny, H., Scott, P. and Gibbons, M. (2001) *Rethinking Science: Knowledge and the Public in an Age of Uncertainty*. Cambridge: Polity Press.

Rittel, H. and Webber, M. (1973) 'Dilemmas in a General Theory of Planning', *Policy Sciences*, 4, pp. 155–169.

Steffen, W., Grinevald, J., Crutzen, P. and McNeil, J. (2011) 'The Anthropocene: Conceptual and Historical Perspectives', *Philosophical Transactions of the Royal Society*, A 369, pp. 842–867, 483.

Wordsworth, W. (2008) *Wordsworth: Poems.* P. Washington (ed.). New York: Random House.

2

Collective learning

Joining the dots

Valerie A. Brown, Judith A. Lambert and John A. Harris

A workshop programme on whole-of-community change produced some surprising results. The Local Sustainability Project (LSP) program, 1990–2015, was delivered against the background of a time of technical brilliance, on the one hand, and disruption and unrest, on the other. Since events such as the Arab Spring and the global financial crisis, social structures have become more unstable and environments further degraded. Information technologies are increasingly reshaping relationships among institutions and nations and creating new relationships between individuals and their society. Most of these changes bring wicked problems; that is, problems that are part of the society that has generated them and that the society has not been able to resolve.

The solutions could lie with independent and open-minded individuals, who can be inside a society asking critical questions and outside it, finding fresh answers at the same time. During the LSP workshop programme we, the authors of this chapter and facilitators of the workshops, found some unexpected relationships. The first workshops, based on current issues, were routine and the results were as expected by the surrounding society. A second workshop series, based on the social knowledge bases of the participants, generated new solutions.

A third workshop series, based on the contributions of independent individuals, identified the multiple dimensions of their minds.

Over 300 workshops were held in Australia, Canada, Europe and the Pacific, 1990–2015, by the Local Sustainability Project of the Australian National University. We formed a workshop team with a leader (Brown), facilitator (Lambert) and observer/recorder (Harris). We had a range of experience: Brown in social change, Lambert with organizational change and Harris in environmental studies. The fourth member of the workshop team, an administrator responsible for the practical arrangements, was drawn from each workshop site. Our team supported workshops initiated by national and local levels of government, communities, organizations and concerned individuals. The communities we worked with differed widely across cultures and interests.

The relationship between individuals and their community has always been a matter for debate. We have Plato's *Republic* and Thomas More's *Utopia*. Thoreau's classic ideal of a humble life by a small pond is very different from Jane Jacob's responsible citizen in a liveable city. An Aboriginal Australian's identity lies in the animal allotted to them at their birth. For individuals in a Hindu community, their cycle of reincarnation may take them from being an ant to becoming a priest. In a beehive, the many highly specialized roles can be exchanged whenever circumstances change. In other words, every individual has their own special relationship with their community.

The Local Sustainability Project workshops built on the work of many influential educators: Michel Piaget's (1951) stages of intellectual development; Paolo Freire's emphasis on the need for learning through conscious reflection (1996); Kurt Lewin's insights into organizational cultures (1948); Carl Jung's 1964 assertion that learning results from people's preferred ways of adapting to the world and John Dewey's commitment to civil society (1910). These provided a powerful background against which to design what we called 'collective social learning' workshops.

The learning framework chosen by the team was David Kolb's experiential learning cycle (Figure 2.1). Kolb had developed the cycle based on extensive experience of individuals in organizations and in government. The Local Sustainability Project team developed it further to include a wide range of individuals and of groups. The design of the collective learning cycle was to bring together all interested parties in the issue, have them agree on an *ideal* solution and review existing *factual* information; then brainstorm *new ideas* and agree on appropriate *action*. After evaluating the action, a fresh cycle moves on to a further ideal, leading to a continuing spiral of ideals, facts, ideas and action. The workshops struck several roadblocks during the early use of the social learning cycle. At the first step of each cycle, sharing ideals, diverse participants could find it difficult to agree on a common goal. Failing to agree generated disagreements that rumbled through the rest of the workshop.

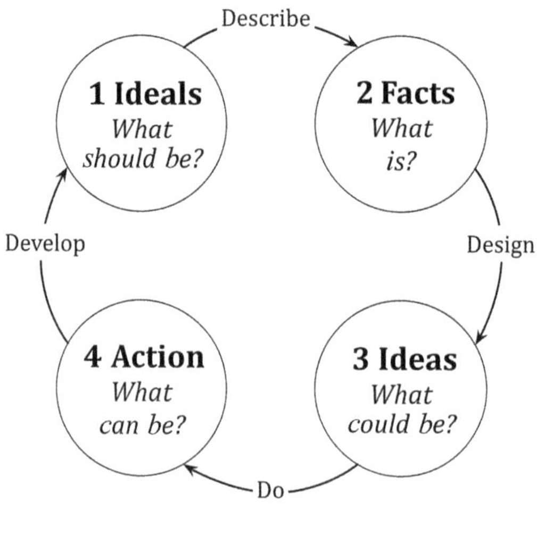

Figure 2.1 **Collective learning cycle (after Kolb 1984; Brown 2008)**

At the second step, finding the facts, differences between those committed to quantitative and qualitative evidence could lead to outright conflict. The third step of the cycle, brainstorming new ideas, could rest on an uneasy base of lack of trust, inherited from stages one and two. In the fourth step, action, suggestions for implementing the ideas were more likely to come from established groups using established methods, unless the project team actively encouraged innovative teams with new ideas.

In developing the steps of collective social learning, it became clear that it is essential to follow the four stages in a definite order (Figure 2.1). It had become routine to follow the scientific era custom of always starting with 'the facts'. However, starting with strongly felt ideals on *what should be* ensures that the collective learning is driven by a desire for change. The next stage, *what is*, establishes the range of facts that allows for the opportunities for, and blocks to, change in the light of the ideals. Next, the participants needed to brainstorm to generate new ideas, *what could be*. For the final stage of each collective learning cycle, *what can be*, what is practical to do, it will be essential that a collaboration is built on learning gained from the learning cycle, rather than perpetuating old divisions. The experience of the project team confirms Kolb's insistence that no learning occurs without the final step to action.

The first series of Local Sustainability Project workshops brought together all the stakeholders to work on a shared wicked problem. The hope was that the stakeholders would form a collaborative team (see Box 2.1). At the time, this was the standard way of responding to wicked problems. It was expected that the stakeholders would reconcile their differences and eventually achieve consensus on the actions to be taken. After several workshops, our small team shared our learning from Series 1.

We found that the word 'stakeholders' implied established interests. This had led to the exclusion of those regarded as secondary stakeholders, such as the next generation, and proponents of aesthetic and heritage traditions. We decided to replace the word 'stakeholder' with 'interests' and invite all interests as primary interests in their own right.

BOX 2.1 COMBINING INTEREST GROUPS: A HEALTH STRATEGY FOR THE ACT

A community organisation received a grant from the Health Department of the Australian Capital Territory to develop a community-wide strategy on sustainability and health for the ACT. Thirty-five stakeholders in the issue were invited to a collective social learning workshop on this theme and 17 accepted. These included Action buses, ANU Green, Canberra Art School, a Credit Union, Landcare, Rotary, YMCA, SHINE (Sustainability and Health in Education), University of Canberra, two commercial interests and two indigenous health organisations. Twelve projects were initiated, with nine proceeding to the final stage of collective action.

The first group workshop brainstormed the following project ideas: recycled grey water, health pathways for indigenous people, nutrition for young people, sustainable consumption research, ambassadors for sustainability, an urban environmental learning precinct, an environmental focus for business and theatre in schools. A second collective social learning cycle with the same group moved on to action. Their proposals came from individual organisations and their membership. There were no proposals for collaborations between the proposing organisations, although there were between their sub-groups.

The projects taken together could have formed a whole-of-community initiative, the aim of the programme. However, only one project suggested this; the final projects remained separate initiatives. They were a Youth Film Festival, Community Gardens, a commercial firm using grey water, Belconnen community action, an indigenous-designed wellness centre, Action buses as a vehicle for creative drama on sustainability, YMCA Bush Capital Lodge, Factor of ten: art display and community activities.

Comment: The theme of the workshops was set by the Health Department, so those who came were expressing an interest in a pre-existing theme, rather than looking for new ideas. There were many imaginative ideas in any case. There was a developmental sequence throughout the workshop process, moving from establishing a set of interests (invitations) to brainstorming open ideas (workshop 1), and then grounding those ideas in action (workshop 2). The considerable difference between the proposals between workshops 2 and 3 seemed to represent the degree of commitment by the proposers. The ideas that came through to action was mainly due to championship from an established organisation, rather than new knowledge or individual ideas.

Our team found that diverse interests often came with a history of conflict. The standard workshop expected to resolve the conflict and enable consensus among the stakeholders. Here again, our team decided that we disagreed with this standard practice of seeking consensus. We had found it more fruitful to bring the conflicts into the open, and so have workshop members understand one another's issues. Attempts to achieve a single answer had been counterproductive, since it set members to competing with one another. We decided to abandon the search for consensus and magnify the diversity. At the time, this seemed like a considerable risk.

In the discussion, we found that many of the participants had never actually listened to others. When differences were respected, when individuals could speak uninterrupted of what they cared about, and only questions of explanation could be asked, collaboration emerged spontaneously. The energy from conflict was redirected to mutual listening and learning (Box 2.1). It is noteworthy that Figure 2.1 contains both boxes and arrows. The boxes of each learning step of ideals, facts, ideals and action contain material generated in the workshop on the chosen topic. The arrows in Figure 2.1 represent the links between each of the stages. The arrows are often ignored in studies of social change and yet they are the essence of bringing about change. In the Local Sustainability Project, the links between the steps in the learning cycle were taken seriously. Focused activities allowed each table of participants to summarize and share their conclusions at each step.

In the first series of workshops, participants were seated around tables whose nine to 11 members were drawn from the range of interests in a topic chosen by the Local Sustainability Project team. Each learning cycle was intended to end in a joint action. Topics ranged from citizen concern for the environment, a city with issues for its environmental health, development of a local government region, national environmental sustainability, transformational change and many more. We did have some success in achieving change. However, the negative experiences of defending open-ended discussions on the issues and being expected to achieve consensus led to much soul searching among team members. They decided to persevere with a second round of the social learning cycle, based on the learning from the initial set of workshops.

In this second round, the selection of the workshop theme moved from being a decision made by the Project team, to the workshop members selecting one that was a shared concern. For instance, in a drought-stricken community, the project team and the workshop participants decided their aims could be summarized under one theme: 'a blueprint for survival'. The shared theme allowed all interests to take an equal part in the discussion, creating a more collegiate atmosphere among the interests of landholders, business and government. A second change was in ways of determining 'the facts' underlying the issues. Instead of assuming facts were all biophysical measures, facts were interpreted as biophysical and social factors hindering or supporting their chosen goal. Everyone could then contribute their own version of the facts.

Most important of all, the second series of LSP workshops was still based on a selected issue. Another workshop dynamic appeared. Between each of the steps of the cycle, groups shared their learning through reporting to the full audience. To everyone's surprise, responses that agreed with each other came from across, as well as from within, the tables set up around interests in the issue. The cross-table groups shared similar aims and objectives. In seeking facts, group members divided into either qualitative or quantitative options. In brainstorming ideas, people found that some were linking ideas across the tables. In forming action groups, the same people wanted to work together.

After much consideration of what, exactly, was happening, our team concluded that the shared positions were coming from different knowledge bases. Tables that had been set up to bring different interests together (learning from difference). However, we found that individuals formed a group with other individuals, community members with other community members, specialists with other specialists, organizations with other organizations and holist thinkers with other holists. Our team identified the groups as drawing on individual, community, specialized, organizational and holistic forms of knowledge (see Figure 2.2).

The findings from the second series of workshops were just as surprising as those from the earlier series. We found that each of the knowledge groups had its own aims, language, accepted content and approved practitioners, making each knowledge group a culture of its own. Once the knowledge cultures were made the basis for invitations to the workshops, the project team found other points of agreement. Each of the knowledge groups had a different method for collecting evidence. Of the knowledge groups, individuals called on their own

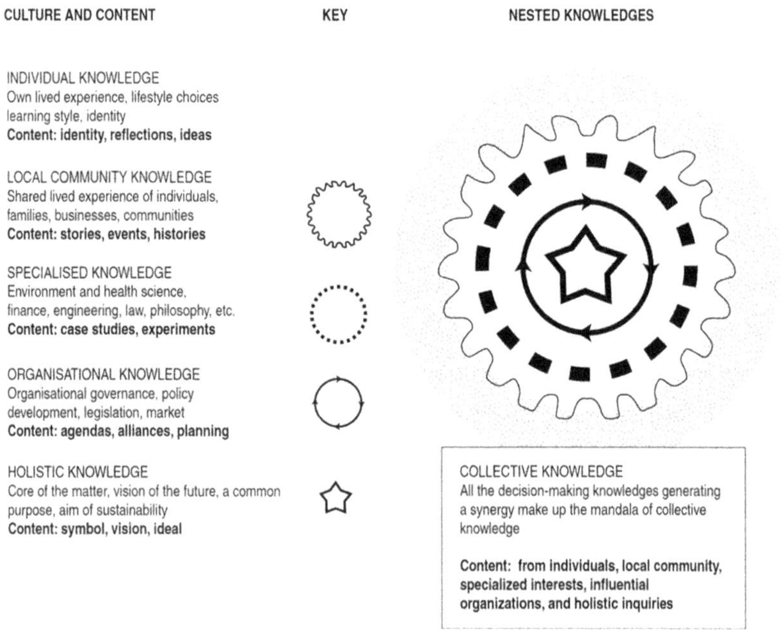

Figure 2.2 **Knowledge cultures of Western societies (Brown, 2010)**

experience; a community drew on shared events; specialists relied on their skilled observations; organizations kept to their own agenda; and holists sought a combination of the knowledge compartments. As part of maintaining the strength of their compartments, each group tended to belittle the others. Individuals were described as biased, communities as self-centred, experts as talking jargon, organizations as self-serving and holists as unrealistic. The accusations were accurate enough to stick. The differences between the knowledge-based groups were distinct enough to represent different cultures of knowledge.

In response to these events, we made more radical changes in the second cycle. We changed the invitation list to members from each of the knowledge groups, who also covered the range of interests. This time, it was workshop participants who chose the focus question. The task was no longer to arrive at consensus, rather to arrive at a mutual understanding of one another's positions. The membership at each table was a mix of participants from each of individual, community, specialized, organizational and holistic knowledge bases.

There was a more positive side to these knowledge compartments to consider. During the scientific era, the knowledge cultures of Western societies have had a strong track record for advancing knowledge. Individuals have been awarded Nobel Prizes for breakthroughs in a wide range of disciplines including physics, medicine, astronomy and literature.

Communities have developed lifestyles that include material possessions that would have been regarded as unimaginable luxuries for most of human history. Specialists have become more and more technically skilled in their specialization, expanding the knowledge base of all disciplines and professions. Organizations have become both powerful and efficient, serving the entire world such as Google and Amazon. Holistic thinkers such as Mahatma Gandhi in India and Jan Smuts in South Africa provided a breakthrough for their societies. In designing a third workshop series to take account of the learning from the second, we were concerned not to throw the baby out with the bathwater. While much was lost by trying to tackle wicked problems by drawing on only one type of knowledge, there was much to gain in bringing the knowledge cultures together in multidisciplinary teams.

The second workshop series had confirmed that Western societies were fragmented into a set of strongly bounded silos of knowledge (see Box 2.2 and Figure 2.2). The boundaries to these knowledge cultures were so strong that their members clung to their positions even in the face of contradictory evidence. In this second series of workshops, we found that the set of socially determined knowledges had formed a hierarchy in which specialized knowledge vied with organizational knowledge for supremacy.

BOX 2.2 COMBINING KNOWLEDGE CULTURES: AN EIGHT-MEMBER ALLIANCE

An eight-member alliance included two councils, two New South Wales government departments (organisations), one academic faculty (specialists) a funding agency (ethics), a regional association of councils (community) and a project management team (ideally, sympathy). The alliance was formed when five participants had put in a similar proposal to the granting body, who asked them to form one collaborative team. They agreed and the alliance received all the grant. After a year, the various interests had not been able to reach agreement on the project design and a facilitation team was appointed to move the project on.

The Local Sustainability Project team designed a change process that included a workshop, a residential weekend and then a second workshop. The aims, agreed to by the alliance, were to facilitate and report back on shared knowledge and capacity building, and to establish integration and capacity building over the life of the programme. The rules for the alliance project were: these are the right people, this is the right time, this is the right place, whatever happens is the right thing to happen and, when it's over, it's over.

The partners in the alliance included members from each of the knowledge cultures: namely, individual, local community, specialist, organisational and holistic. A synergy occurred whenever there was equal contribution from each of the knowledge bases. The workshop moved through a series of events that brought all the knowledge cultures into play for all members of the alliance: role plays, diaries, dialogue and a multiple knowledges grid.

Comment: There were different workshop contributions when participants were acting as a representative of their organisation and as a member of a knowledge culture. A multiple knowledges grid was developed that crossreferenced the allotted tasks and members of the knowledge cultures. Concluding comments from the retreat participants indicate a strengthening of mutual trust, understanding of and respect for different ways of knowing and differing value sets within the alliance. This provided a sound basis from which the alliance project could further its achievements in their teamwork beyond the individual knowledge cultures.

During the second workshop series, based on the knowledge cultures identified in Workshop 1, we, the workshop facilitators, noticed that there were independent individuals who took up a range of positions. Sometimes an individual took the position of one of the knowledge cultures; sometimes they took a fixed position of their own. Sometimes an individual took up an independent position, asking critical questions at every step of the cycle and consulting with others on possible answers. This independence confirmed the proposition with which this chapter began: the influential role of an independent individual in difficult decisions. However, there still remained controls on their freedom by the knowledge cultures.

The research team designed a workshop with individuals working in partnership with knowledge cultures in the hope of finding common ground. This third workshop series was based on individual difference, rather than on interest in the issues (Series 1) or knowledge cultures (Series 2). Workshop members in this series (Series 3) were asked to participate as individuals speaking on their own behalf; not as a stakeholder in the chosen issue or as a member of one of the knowledge cultures. As individuals, they were asked to draw freely on other sources of information, from interests in the issue and knowledge cultures and to combine that learning with their own direct experience.

After several of the workshops with emphasis on independent individuals had been held, another distinctive pattern emerged. The distinctive group in this third series, based on the strong individuals, was the members who drew on the multiple dimensions of their minds. Through our own observations and after consultation with workshop participants, the dimensions of their thinking were identified as:

Biophysical – the physical environment in which an issue is set. Arrived at by observations, measurement and formal reports.

Socioeconomic – the social environment, including cultural rules and the socioeconomic system (a prevailing emphasis in Western culture). Arrived at through personal involvement, and the narratives and records of the community.

Ethical – the principles governing relationships between individuals and between individuals and society and between individuals and their environment. Arrived at through reflecting on a cultural framework and/or a personal commitment to a way of life or a religion.

Artistic – sensitivity to the patterns in natural and in social systems, arising from the capacity for inspiration within each human being. Arrived at by both expressing, and rebelling against, cultural norms.

Sympathetic – recognizing a shared understanding with another human being or group or with another species. Arrived at through feelings of openness, trust and shared experience.

The dimensions of mind are represented in an account of a workshop (see Box 2.3) and as diagram (see Figure 2.3).

BOX 2.3 COMBINING MULTIDIMENSIONAL MINDS: HUMAN ECOLOGY WORKSHOP

In a workshop with a group of independent individuals at the Australian National University, the focus was on the idea of sustainability in the Anthropocene. The request from the workshop facilitator to participants was to explore their interest through each of the multiple dimensions of their minds: biophysical, socioeconomic, ethical, aesthetic and sympathetic. The participants were already involved in transdisciplinary thinking.

(Continued)

Participants included three physical scientists, two social scientists, one musician, one environmental activist, a statistician, an administrator, two PhD students and a planner with local government.

Responses from two of the scientists were that only the biophysical was worth including in a research project. The other scientist included all the dimensions and asked the question, how can each of these be subject to proof? One social scientist grasped all the dimensions as a method for her research project in public health. The other questioned the need for both socioeconomic and sympathetic, considering that they belonged in the same category.

The musician recognised all dimensions with enthusiasm. The environmental activist was able to recognise each dimension in their campaigns, but said they would only take account of the first two in an action. The statistician welcomed the full set of dimensions as a guide for collecting data. The administrator was surprised at the way the dimensions opened up her field of responsibility; more important to her was the use of them in understanding herself. One PhD student argued that the aesthetic had no place in her thesis topic. The other said that she would include the dimensions in a special chapter, since examiners were unlikely to understand them. The local government planner said that they could identify all dimensions in their own planning regime although they would not tell other planners.

Comment: This was a particular sample of independent thinkers, since all participants were already interested in transdisciplinary thinking. Five of the 12 members had limitations on their recognition of all of the dimensions of mind, mostly due to their existing knowledge culture. The other seven could recognise all dimensions and would use them in some capacity. The results from the workshop confirmed the proposition that not everyone can recognise all the multiple dimensions of their mind. Those that could were keen to apply them in their existing activities, although they thought that this would be difficult for their colleagues to accept.

In Figure 2.3, the multiple dimensions of mind are placed within their context. The dimensions make up one undivided mind. In using our minds, we each reflect on a particular context through our personal reflections. The reflections can be outward, on the social and physical environments in which we find themselves. We can reflect inwards (usually called introspection) identifying our own assumptions, memories, hopes and fears. The sum of the reflections gives us an idea of our own identity and position in the world.

In the LSP workshops, independent individuals' basis for their thinking differed significantly from that of the

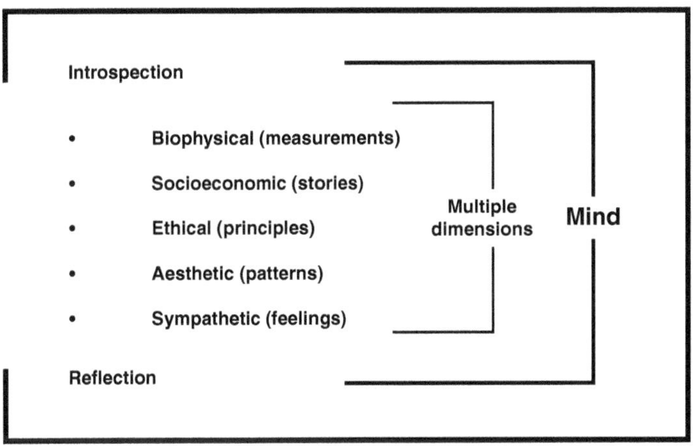

Figure 2.3
The dimensions of individual minds in context (after Hocking, Brown & Harris 2016)

socially determined knowledge cultures. While the latter tended to look for authorities in their own field, the individuals drew on their own experience and a wide range of different sources. Individuals were more likely to make their contributions from the interconnected biophysical, socioeconomic, ethical, artistic and sympathetic dimensions of their minds (see Table 2.1). Individuals also interpreted the focus question for the workshop in their own way, thus creating a fresh dialogue. In the previous workshops, members of knowledge culture and stakeholder workshops were more likely to keep to their pre-existing focus.

There are some things every individual has in common. One is the capacity to learn. Whatever the relationship between an individual and their society, physicist-philosopher David Bohm points out that living with change necessarily involves learning. He adds that learning comes from difference, not more of the same. Cultural diversity can lead towards, not away from, effective responses to local and global issues, bringing fresh ideas and new options. National reputations can depend on the gifts of individuals. Minimalist Japanese designers, creative French artists

Table 2.1 Comparing the multiple dimensions of individual minds with the socially constructed knowledge cultures

Dimensions of mind	*Reflection*	*Knowledge cultures*
Biophysical	Experience, reflections	Measurements, observations
Socioeconomic	Whole-of-community	Specified groups
Ethical	Personal ethics	Cultural rules
Aesthetic	Patterns	Fashion, design, art
Sympathetic	With others	Own group

and efficient British administrators come to mind. The individuals who made up the 'huddled masses' that reached New York at the turn of the last century turned the United States into the most powerful nation in the world.

On the other hand, the voice of an individual can be missing altogether. They may not have been considered important enough to be listened to, or be seen as too much of a rebel, to be allowed into the mainstream. Yet the independent individual is the one most likely to stand outside the power plays that block any action on serious issues. They are likely to have different ideas and proffer different solutions to a long-established problem. But before they can be heard, they must have access to where decisions are made, and the freedom to speak their mind.

The sweeps of history confirm that independent individual voices are at risk of being silenced by their society: Socrates in ancient Greece, Joan of Arc in Medieval France, as well as Martin Luther King in the USA and Mahatma Gandhi, closer to our own times. Hercules had protected himself from his own society. To compound the problem, members of Western societies live with a series of paradoxes. In one paradox (two things that cannot both be true at the same time), I am called on to be ambitious, to win at all costs and to grasp every opportunity to increase my power. I am also told to be modest, to be a good loser and to be a team player. I am expected to plot for power and be a humble citizen at the same time. It is no wonder we can find ourselves confused about our own identity.

Another paradox has emerged from the scientific era. As a citizen, I am required to be impartial, objective and focus on selected parts of an issue. In my family, I am being told to be sympathetic to others, to be creative and to focus on the whole. It appears to be an impossible agenda for one person to display objectivity and sensitivity as one. Yet all of us are faced with these apparent oppositions, being a leader and a follower, and drawing on facts and on feelings, all at once. Yet we all do all of that all the time.

Even René Descartes, so often blamed for the scientific tradition of allowing only logical argument, recognized the worth of different approaches to knowledge: 'And that the diversity of our opinions does not arise from some being endowed with a greater share of reason than others, but solely from this, that we conduct our thoughts along different ways, and do not fix our attention on the same objects.' In the Local Sustainability Project study of changing communities, individuals were indeed found to conduct their thoughts along different ways and fix their attention on different objects, two avenues of the learning from difference sought by David Bohm.

From workshop Series two and three, it was possible to discern two streams of thinking, one derived from socially determined groups (Box 3.2) and the other from independent individuals (Box 2.3). Not surprisingly, there were connections between the two, given the same individuals could speak for themselves and lived in the same society. Both workshop series confirmed the influence of the longstanding dominance of biophysical sources of evidence in both groups. The conflict that erupted when the groups were responding to the 'facts' step in the learning cycle was partly due to a collision course between the members of the knowledge cultures. Yet effective solutions to complex problems require the contributions of the knowledge cultures and individual minds. Relationships between the open individual and the closed knowledge groups are listed in Table 2.1.

Series three workshops identified the danger that individuals could be swamped by the dominance of society, and their society swamped by the strong positions of individuals. In the Local Sustainability Project, the differences between individuals drawing on their own mind and drawing on socially constructed ways of thinking were not treated as opposites. The aim was to develop a relationship in which society and individuals negotiated opportunities to learn from one another, as David Bohm would recommend.

In the third workshop series, personal experience was given full weight. The simpler categories of physical and social proved to be insufficient to take account of the complexity. The physical and the social categories expanded to the biophysical and socioeconomic, recognizing the critical state of the biological component and the dominance of the economic. The ethical dimension of the members' minds was fulfilled by the ideal of critical loyalty. Members had agreed to support the other members in meeting their own goals and had also agreed that they had a responsibility to share constructive criticism. Artistic sensitivity was the dimension least acknowledged, which may reflect the Western culture. One of the most powerful dimensions was the allocation of sympathy. In the socially determined knowledge groups, sympathy with members of one's own group was often stronger than sympathy with another not of one's own group.

These Local Sustainability Project studies and their findings were not confined to the statistical and measurement tradition. For every individual, all dimensions of their mind could become actively involved. This active involvement was not to be taken for granted. Box 2.3 contains profiles of the ways in which individuals may stick to their favoured knowledge culture. The participants could ignore their own dimensions of thinking; acknowledge only one dimension and ignore the others; and all could recall their delight in drawing on all dimensions. The conclusion of the

Local Sustainability Project team was that it was essential to combine group and individual thinking when responding to the wicked problems of social and environmental change. Case studies of collective social learning cycles, as in the examples in Boxes 2.1, 2.2 and 2.3 identify the contributions from interest groups, knowledge cultures and independent individuals at each step of the learning cycle.

Conclusions

Relationships between learning, knowledge, power and identity proved to be so intertwined that a separate interpretation of their relationship was needed for each learning cycle. It was important to 'ground truth' a collective learning cycle; that is, follow each cycle to completion in its own environment. Putting a learning cycle into action maximizes the collective capacity to act on wicked problems; joining the dots, as it were.

In this chapter, we have explored the evidence collected by the Local Sustainability Project on individual and socially determined learning cycles in relation to the wicked problems of the Anthropocene. The emphasis throughout has been on independent individual thinking and the importance of welcoming diversity. Currently, there is a practice of clinging to one way of thinking from among the possible ways of thinking, a practice that both impedes the capacity for individuals to use all of their mind and bars the way of working together.

Resolving complex problems requires strategies that link the complexity of the issues, socially constructed forms of knowledge, and the biophysical and social and ethical and artistic and sympathetic dimensions of individual minds. Chapter 3 contains a description of each dimension of mind in practice and of the ways in which the dimensions act, not in isolation, rather in unison.

Bibliography

Bloom, Allan (1968/1991) *The Republic of Plato*. New York: Basic Books.

Bohm, D. (1996) *On Dialogue*. London: Routledge.

Bohm, D. (1994) *Thought as a System*. London and New York: Routledge.

Brown, V. A. (2015) 'Utopian Thinking and the Collective Mind: Beyond transdisciplinarity', *Futures Journal*, Vol. 65 *Advances in Transdisciplinarity 2004–2014*, 209–216.

Brown, V. A. (2010) 'Collective Inquiry and its Wicked Problems' in Brown, V. A., Harris, J. A. and Russell J. Y. (2010) (eds.) *Tackling Wicked Problems: Through the Transdisciplinary Imagination*. London: Earthscan, pp. 4–81.

Brown, V. A. (2008) *Leonardo's Vision: A Guide to Collective Thinking and Action*. Rotterdam: Sense.

Brown, V. A. and Harris, J. A. (2014) *The Human Capacity for Transformational Change: Harnessing the Collective Mind*. London: Routledge.

Brown V. A. and Lambert J. A. (2013) *Collective Learning for Transformational Change: A guide to collaborative action*. London: Routledge-Earthscan.

Descartes, R. (1637/1955) 'Discourse on Method', in E. S. Haldane and C. R. T Ross (eds.) *Philosophical Works of Descartes*, 2 Vols. New York: Dover, Part 1, p. ii.

Dewey, J. (1910) *How We Think*. Boston: Heath.

Freire, P. (1970/1996) *Pedagogy of the Oppressed*. Harmondsworth: Penguin.

Hocking, V. T., Brown, V. A. and Harris, J. A. (2016) 'Tackling Wicked Problems Through Collective Design', *Intelligent Buildings International*, 8(1), 24–36.

Jung, C. (1964) *Man and his Symbols*. New York: Doubleday.

Kolb, D. A. (1984) *Experiential Learning: Experience as a Source of Learning and Development*. Englewood Cliffs, NJ: Prentice-Hall.

Kolb, D. A., Lublin, S. and Spoth, J. (1986) 'Strategic Management Development: Using Experiential Learning Theory to Assess and Develop Managerial Competencies', *Journal of Management Development*, 5(3), 13–24.

Lewin, K. (1997) *Resolving Social Conflicts & Field Theory in Social Science*. Washington, DC: American Psychological Re-published Association.

Lewin K. (1948) 'Defining the Field at a Given Time', *Psychological Review*, 50, 292–310.

More, T. (1530/1995) *Utopia* (trs. Logan, G. M., Adams, R. M. and Miller, C.), Cambridge: Cambridge University Press.

Piaget, J. (1951) *The Child's Conception of the World*. London: Routledge & Kegan Paul.

3

Multiple dimensions of mind

Parts and wholes

Valerie A. Brown

Chapter 3 contains the story of how a research team of which I was a member discovered that independent individuals drew on the biophysical, socioeconomic, ethical, aesthetic and sympathetic dimensions of their minds. The dimensions did not stand alone; they were integral parts of each individual mind. As the study continued, the team found that all dimensions were available to every individual, although not everyone recognized all of them. Most of the individuals had favoured entry points to their thinking, usually the physical and social dimensions of their lives. The next step is to explore the significance of these dimensions of mind for an individual in everyday life.

Three generations have passed since Alvin Toffler's 1970 *Future Shock* was a runaway best seller. His argument was that from then on, each generation would live in a world quite different from that of their parents. After a longstanding tradition of expecting children to follow in their parents' footsteps, this was indeed a shock. Since the 1970s, generational change has become taken for granted. Each generation has had to work things out for itself. Thinking only what my current society is telling me to think is no longer a safe option. Things are sure to change. To understand the era that I live in, I need access to every one of the multiple dimensions of my mind.

In living with continual change, I feel the need to ask: 'What does this mean for me?' The answer will involve the multiple dimensions of my mind and so my identity. A simple example: once local news arrived in a newspaper I had ordered from a newsagent, thrown from a delivery van into my garden every morning. Twenty years later, news on world events now arrives uninvited into a television set right inside my home. The *physical* channel for the news is the television set. Watching the news with my family reveals my *social* lifestyle. The programme topic, perhaps starving children in Ethiopia, alerts my *ethical* sense of the need for justice. The *artistic* design of the advertisements helps capture my attention. Television news of starvation is detached from my life, so it may reduce or heighten my *sympathy* for the afflicted.

In reading the local paper, I had learnt how to become an active citizen. From the new medium that Marshall McLuhan called the mechanical bride, I had unwittingly become interested in international affairs. In other words, the apparently simple shift from newspaper to television changed who I think I am. In such a constantly changing world, I find myself wondering where I fit in. My experience of change is usually interpreted in the ways that my society expects: proper behaviour for my gender, income, age, ethnicity and education. Yet when I encounter change, I am a whole person, I think along multiple dimensions and my experiences are mine alone. As an adult the multiple dimensions of my mind are already in place. I already access the biophysical world as an ecologist, the social world as a citizen, the ethical as a citizen, the aesthetic as a homemaker and the sympathetic as a parent and grandparent, all within the one undivided mind. I may or I may not fit into the social expectations of me in all these roles. In this 21st century, I am hoping for less social control and greater respect for independent thinking; not just for me, for everyone.

How adult human beings think about things has been the focus of many different studies. The summary in Table 3.1 adds multiple intelligences, coloured hats, knowledge cultures and the traditional disciplines to multiple dimensions of mind. While the focus on individual dimensions of mind is not the whole story, it is a *crucial part* of the story. Here we are concentrating on the ways that each individual can think for themselves, remembering that the human mind can handle this complexity. Howard Gardner's multiple intelligences are discussed later in this chapter.

Before going on to consider the strengths of the multidimensional mind, we should look at the barriers lying in the way. The following proposition captures the helplessness of individuals who have not learnt to think for themselves: 'The leaders have no need to seize the rights of the citizenry. Rather the citizenry, infused with fear, will offer up all their rights unto the leader, and do it gladly so' aptly captures the essence of power relationship between individuals and their society. People will follow authority simply because it *is* authority. The authority allows

Table 3.1 Qualities of mind, by author

*Intelligences**	*Thinking hats**	*Dimensions of mind*	*Knowledge cultures*	*Disciplines**
H. Gardner	*E. de Bono*	*V. A. Brown & J. A. Harris*	*V. A. Brown*	*S. Toulmin*
Intrapersonal	Feelings	Biophysical	Individual	Philosophy
Interpersonal	Information	Socioeconomic	Community	History
Naturalist	Judgement	Ethical	Specialized	Mathematics
Musical	Optimism	Aesthetic	Organizational	Biology
Logical	Creativity	Sympathetic	Holistic	Chemistry
Linguistic	Thinking			Physics
Existential				Medicine
Spatial				

*In no particular order

them to feel safe and so there is no need to resist it. Under this scenario, citizens lose their right to even their own identity.

Each of the knowledge cultures (individual, community, specialized, organizational and holistic), identified in Chapter 2, zealously guard their knowledge territories. Relying solely on these socially determined knowledges can effectively block access to all multiple dimensions of your mind. A physical scientist considers subjective evidence from an individual contaminates their careful measurements. To make expert knowledge available to everyone challenges a specialist's social authority. An ethical community service may ignore the diverse needs of individual clients. An artist may be treated as an outsider in their community, so their independent voice goes unheard. Someone different from yourself may be considered 'the other', which can mean that you receive little sympathy. Without your own independence, all dimensions of your mind can be unavailable.

Individuals who focus only on their own mind and their own interest have been disastrous for themselves and their communities. At one extreme, a dictator or a terrorist will focus only on himself and the outcome he has chosen. At the other extreme, the single-minded individual becomes alienated from her society. Both an autocrat and a loner are inwardly focused, incapable of following any path but their own. Large and small injustices arise from this self-interested mono-dimensional thinking, from the Holocaust to the minor regulations of officialdom.

The 20th-century cult of individualism creates a distortion of the independent mind. Followers of the cult say, 'What's in it for me?' and 'My prime responsibility is to myself'. The irony is that this cult has emerged at the close of a scientific era that rejected subjectivity. Individual members of established professions, cultures and religions tend to reject the thinking of non-members and privilege those that think like themselves. Advertisements offer remedial courses in accessing your personal power for yourself. The risk in such courses is that the individual can be isolated from their society and from their sympathy with others.

The history of dominance of physical over other ways of thinking has led to considerable confusion between brain and mind. My physical brain is not my thinking mind. My brain provides the structure that allows me to think; it does not do the thinking. Although shaped by cultural rules and physical limits, my thinking remains under my own control. In a damaged brain, the sufferer's thinking can be re-allocated to another part of their brain. In an actual case, a 25-year-old accident victim was rescued from a morgue refrigerator with a fifth of her brain destroyed. Two months later she emerged from her coma unable to move or speak. As she told me later, as a trained social worker, in the recesses of her mind she knew that it was possible for her to retrain her brain. So she did. By 15 years later she had given birth to five children and was back in her profession. While the brain allows the mind to function, it is the mind that is in control.

A human mind is a learning and inquiring mind, sensitive to other minds. Every mind holds knowledge about *how* to establish a practice (ride a bicycle, set a broken arm, employ multiple ways of thinking) and knowledge *that* a statement is true (war is inevitable, the universe is expanding, human beings can draw on multiple forms of thought). Often unconsciously, the multiple dimensions of a human mind are accessed as day-to-day practice. While not forgetting that the dimensions of each individual mind make up a seamless whole, it is rewarding to examine each dimension for its own sake. So here is an informal overview of each dimension of our minds and how they link to other dimensions, before turning to ways in which they become a whole:

Biophysical: We all know that the biophysical dimension of our mind is created through the senses: sight, taste, smell, hearing and touch. This now includes extensions of the senses like an electronic microscope and a microphone. We add to this dimension in many practical ways: through observations, physical and thought experiments, statistical calculations and case studies. Your links with the other dimensions can be through fellow practitioners, professional ethics, beautiful experiments and close friends. René Descartes' separation of mind and body allowed the biophysical dimension to be thought of as standing by itself. Yet Descartes was deeply religious, concerned about design and involved in mathematical politics. Einstein was a skilled violinist and a jokester with a small circle of close and sympathetic friends.

The inadequacy of drawing on only one dimension at a time became apparent during the 20th century. Addressing health issues required knowing about the context in which they emerged. Sending astronauts into space without knowing that they were 'the right stuff' was a recipe for disaster. Scientific practice became accepted as needing far more than laboratory skills and objective observations. Looking beyond objective science, Silvio Funtowicz and Jerome Ravetz developed a post-normal science. They went beyond the sole focus on objectivity to a broader and richer science, encompassing all dimensions of mind. This immediately spread through the scientific community as being closer to the way the world actually functioned and people thought. Funtowicz and Ravetz proposed that post-normal science was needed whenever 'facts are uncertain, values in dispute, stakes high and decisions urgent'.

Socioeconomic: The socioeconomic dimension of our minds is concerned with the formal and informal relationships among individuals and between groups. Connections with the other dimensions are strong. C. Wright Mills called the leap of the mind in which all aspects of a society are understood as a whole 'the sociological imagination'. As well as by observation, social relationships are confirmed through listening, feeling and imagining what it might be like. Methods of inquiry into what forms a certain society include reflection, translation, narrative, demographic trends, structured and unstructured observations and searching records.

Emphasis on economic relationships has led to the absorption of economic thinking into the social dimension. This dimension gives access to the richness of language, the capacity to think about thinking and the ability to reinvent the rules. The socioeconomic dimension is concerned with the ways in which humans live in their societies. Government, law, education and economic systems are ways of structuring those societies. Governance can be by democracy, meritocracy, an autocracy and a hierarchy. An emerging contribution to a legal system linking social control with sympathy is restorative justice, a system based on mediation between transgressor, victim and community expectations. Also emerging is a movement towards regenerative practices, the repair and renewal of social and natural systems.

Ethics: Personal ethics are seldom openly discussed, although they are the values and moral principles that guide all our relationships with others. Every human being, other than a psychopath, holds an ethical position that guides them on whether their behaviour is good and bad, right and wrong. Moral choices are often confused with biophysical quantities (right and wrong answers) and artistic integrity (beautiful and ugly). The principles that one group develops may be poison to another group and so cause conflict and debate. Different ethical positions can inspire, change, conflict and confront. Moral choices can alter with their time and place. On a Christmas Eve in Flanders during the First World War, with the greatest loss of life of any war in history, the men in the trenches heard singing. Slowly men emerged from both sides of the trenches and sang Christmas carols together. The next day, the shooting continued. Yes, ethics are situation dependent. One person might consider the principle of always telling the truth an absolute value, while another might find it too painful to practice. Isaiah Berlin argues that even when the values of two people clash, it does not follow that one is true and the other false.

Aesthetics: The artistic dimension of the individual mind finds ways of expressing the human spirit. It has been described as a rhythm more powerful than the rhythm of human reasoning. The poet John Keats concluded that: 'Beauty is truth, and truth, beauty. That is all you know on Earth, and all you need to know.' Keats' all-pervading sense of the artistic has competing interpretations. For some, art is an aspect of culture associated with the elite and the privileged. This goes with the belief that great artists can only be appreciated by those with the sensitivity to understand them. In contrast, others believe that any of the arts can provide a form of self-expression as a crucial part of life. A difficulty is the restriction of the word 'art' to the visual arts. As a dimension of mind, the word artistic encompasses all forms of human expression. Dance, poetry, music, storytelling, icons, symbols, self-discovery, experiments, mathematics, dreams, visions and more – anything that is open to a judgement of being beautiful and ugly.

The difference between the logical and the creative mind is illustrated by my experience in lands lived in by the Australian Aboriginals for 40,000 years. Powerful icons called Wandjinas are painted in ochre on rock walls. Guided there by an Aboriginal, I said how wonderful to see the paintings of the Wandjinas. He said firmly: 'That is not a painting of a Wandjina; it *is* a Wandjina.' My literal mind had failed to match the creative imagination of a people who had fully understood they had called into being the Wandjina itself.

Sympathy: Sympathy is a special case of communication between individuals, described by the philosopher Martin Buber. He wrote of the significant differences between sympathy and empathy and between *you* and *thou*. Sympathy is a feeling *for* another; empathy as feeling as if you *are* the other. Once all European languages recognized the distinction between the formality of *you* and the intimacy of *thou*. In English, *thou* has been lost altogether; in line with the trend towards objectivity, German still has *Sie* and *du*; French has *vous* and *tu*.

Sympathy and empathy are feelings that can be communicated without words. Watching a London apartment tower on fire on television, one cannot but feel deep sympathy for people imprisoned with their children on the upper floors. Firemen standing below locked eyes with one parent desperate to get their child out of the burning building. The empathy between each fireman and parent was palpable; it went far beyond sympathy. In those seconds, a mother or father threw their child for a fireman to catch. Not one child was lost. Sadly, some of the parents were.

This account of each of the dimensions of mind has treated each of them as a whole in itself. However, that would be seriously misleading if I did not explore the many ways in which the dimensions are part of an indivisible whole. The whole of our minds is not only greater than the sum of the parts, it is different from it, too. The whole provides the necessary synthesis and synergy from combining the dimensions of mind. Once I recognized the need to find pathways to this synergy, I found then everywhere I looked. I found them in the leaps of the imagination offered by a school of French philosophers of the mid-20th century.

Albert Camus, Jean-Paul Sartre and Martin Heidegger shared a strong sense of the integrity of the mind of an individual. They each explored the ways of thinking as a free individual, rather than a tool of society. Unlike the powerbrokers of their time, these authors supported the idea of an individual as a free and responsible agent. They considered individuals as capable of determining their own development through acts of free will, although in very different ways.

Albert Camus argued that, given the many paradoxes that make up a society, to think that any one individual could bring about change was absurd. Individuals would always be isolated and marginalized by the stronger powers of their society. However, the human spirit and the commitment to being human meant that the individual could not help trying to change their world. Camus believed with Shakespeare that: 'The fault, dear Brutus, lies not in our stars but in ourselves.' The challenge for individuals was to stop feeling and acting like underlings and accept their own capacity to act. Camus's followers were faced with a paradox. His ideas proved depressing enough for individuals to drop out of society and inspiring enough for them to work towards changing it.

Jean-Paul Sartre considered that individuals are self-invented, figments of their own imagination. It follows that there is no reality except for the one that individuals make for themselves. Sartre's work on the freedom of the individual inspired the 'flower people'. They felt free to consume drugs, maintain open relationships and push against social controls. It is not surprising that this led to much social unrest. Sartre's thinking changed automatic acceptance of the dominance of society over the individual, forever.

Martin Heidegger identified humans as the only beings who can *think about their own thinking*. However, if limited to their own experience, individuals have only a partial understanding of their world. Heidegger considered that individuals can only think about the things they have already discovered in their everyday experience. Arriving at an understanding of complex issues requires direct experience of the issues themselves.

Having found these wide-ranging discussions on the dimensions of human minds, I was faced with another problem. With the current emphasis of our society on divisions rather than connections, most psychological and physiological discussions on the human mind are based on a divided mind. I had already concluded that it is as important to understand the integrity of the whole mind as it is to understand the parts. Combining the interpretations of the three philosophers, I learnt of the need to rise above the apparent absurdity of attempting to change my society and still go on trying; to clarify my identity in the relationships between myself and my society; and to draw on all the dimensions of my mind in order to understand my society. However, to absorb new ideas and think for myself is easier said than done. The pull from one's society is very strong. Fortunately, I have been able to find a body of thought and a history of successful examples that confirms that individuals are able to live their lives with personal integrity, drawing on all of their minds.

An example of an individual who successfully drew on all dimensions of her mind is a tracer who worked in a clinic at the time of the then-fatal AIDS virus (acquired immuno-deficiency syndrome) (see Box 3.1).

BOX 3.1 MULTIDIMENSIONAL MIND: INDIVIDUAL

A multidisciplinary AIDS clinic was staffed by medical clinicians, researchers, nurses, administrators and a dedicated AIDS tracer. AIDS is spread by sexual contact, so the tracer's job was to identify the contacts of existing sufferers. Her job was more important than that of the clinicians and nurses, since at that time there was only palliative care. A tracer knew about the physical symptoms of this fatal disease; the social dimension that had criminalised same sex relationships; the ethical principles of the health workers who were willing to treat the disease; and some health practitioners a distaste for the origin of the disease. She had shown had enough sympathy for the sufferers for them to trust her with their contacts.

One particularly conscientious tracer had developed a wide circle of contacts. She noticed that one person from within the AIDS population had not developed AIDS. She advised the other members of the clinic who took little interest; the clinicians were busy prescribing drugs and nurses giving counselling. A year later that person had still not developed AIDS. This time the tracer's suggestion that such an individual could be the source of an immunisation vaccine got through to her colleagues. They checked her story and wrote a widely quoted specialist paper in which the tracer was not mentioned. A feminist outcry led to apologies and acknowledgement. The point here is that, even in a multidisciplinary clinic, only a member who applied all dimensions of her thinking was able to make this crucial discovery. A further point is that single-minded thinkers were not yet ready to acknowledge the value of a multidimensional mind.

The AIDS tracer whose story is told in Box 3.1 could not have given names to the dimensions of her mind on which she was relying. However, she knew full well that she was using them. A step towards drawing on the whole of one's mind is to reflect on the parts themselves as if they were wholes. While this reflection brings a richer and deeper understanding of the whole, it would be a mistake to imagine that the parts exist without the whole. Inside the brain there are innumerable connecting pathways. In practice, the mind is indivisible. As we have seen, even if the brain can ordinarily be mapped by function, if one part is damaged, other parts can take over.

Albert Einstein bequeathed his brain to science. Scientists queued up for a chance to find out whether his brain differed from everyone else's. For 30 years they tried weight, shape, anatomy and DNA patterns. Finally, a scientist discovered a difference. The corpus callosum, where links to the two hemispheres cross over, was not the usual two separate ropes of tissue, but fused into one mass. In other words, Einstein had a more complete synthesis between the two sides of his brain than the rest of us. This gives a possible explanation for his capacity to think of his quantum theory papers, whole, when a mere 18 years old.

As well as drawing on theoretical ideas, there are practical ways to consider the relationships between the multiple dimensions of our minds. Drawing parts into a whole is often compared with a jigsaw puzzle. However, a jigsaw is a predetermined whole, broken up into predetermined parts and then re-assembled. An integrated mind cannot be separated into parts and does not produce the same picture that it started with. Jigsaw pieces have no value except as parts of a whole, while the dimensions of mind all exist in their own right. While the copied picture can be beautiful, it is fixed. It cannot aspire to the synergy that accompanies the dynamic synthesis of the dimensions of mind.

I have found that the idea of a map is more useful than that of a jigsaw. As a map needs three reference points to determine a given area, a map of the mind would take as a reference point all five of its dimensions. The world of a map can be checked by going there, a process called 'ground truthing'. A map of the mind can be checked in a real world, too – it is called experience.

The theme of a map is as important as the contents. The theme predicts the parts. For a multidimensional mind, I can construct a physical map (space, plants and animals, diseases); a socioeconomic map (population flows, resource use, services); a diagram of ethical positions (religions, cultures, cults); a map of artefacts (map design, art works, landscapes); and a map of sympathetic relationships (individual experiences, memories of oneself and/or close friends).

In the 21st century, the advent of information technology has vastly increased the capacity for combining multiple contributions into a whole. Twitter provides maps of the mind in 60 characters, Facebook through symbols for like and dislike, Google maps through invasion of people's privacy and Wikipedia as a compendium of ideas. In Part 2, of this book, we (the editors) have mapped the multiple dimensions of mind through different ways in which the chapter authors have thought of the world. In reading such a map, it is essential to remember that none of the dimensions ever exists alone.

Anthropologists Margaret Mead and Gregory Bateson, botanist Eric Ashby and mathematician John von Neumann, among others, met in the series of Macy Conferences in the 1940s/1950s. The aim was to generate an overarching framework for their emergent systems thinking. The conference was a success to the extent of establishing a school of systems thinking that encompassed all their ideas. From systems thinking came the recognition that all living systems, from a single cell to the planet, were self-organizing. So too is the independent mind.

These are all classic examples of attempting to bring parts into a cohesive whole. Box 3.2 contains a practical example from the everyday world. It is the story of my experience of a committee charged with the responsibility for protecting the World Heritage icon, the Great Barrier Reef. The interactions among the committee members offer insights into the power dynamics among independent individuals, that is, the members of the committee. Some members called on a single dimension of their mind and some on all the dimensions. This difference was echoed in the power relationships in the committee.

BOX 3.2 MULTIDIMENSIONAL MINDS: THE COMMITTEE

In the 1990s the Australian government appointed a committee to oversee the impact of coastal activities on the Australian Great Barrier Reef. The chair was a local grazier. Members came from tourism, finance,

(Continued)

Federal and state governments, conservation, rural industries and science. I was the consumer member, charged with the interests of those dependent on the reef: tourists, small business owners, researchers and future generations.

In our discussions, single-interest committee members tended to draw on incompatible dimensions of their minds to make their case. Tourism interests referred to the aesthetic and the biophysical, saving the beauty of the reef while denying the biophysical damage. Rural industries were seeking profits from both farming and the reef, when runoff from industry was destroying the reef. The ethical standing of the reef's World Heritage status struggled against the political pragmatism of government agencies.

In contrast, other committee members drew on all the dimensions of their minds, some consciously, some unconsciously. As the chair heard from each of the interests on the committee, she broadened her concerns from her own local interests to include those of all committee members. Two of the scientists were designing an evaluation of save-the-reef programmes that included all dimensions, resisting criticisms that this spread their project too thinly. Taking all consumers' interests into account meant consciously drawing on all dimensions of my mind. Drawing on all the dimensions created a common language among the committee members.

The power distribution was uneven across the committee. Single-issue members tended to rely on the dimension closest to their interests, such a government agency to the socioeconomic dimension, a biologist to the biophysical and the conservationist to the ethical. The organisations had the strongest positions, followed by the scientists. Each interest was more likely to accept advice from other single interests than from the holistic thinkers. By the same token, the multidimensional thinking of the chair, the evaluators and consumer interests meant a more complete understanding of an issue and of the other interests on the issue. Even as individuals, this gave them a stronger position from which to influence committee decisions.

In this epoch of the Anthropocene, I still have confidence that independent thinkers who draw on all dimensions of their minds will prevail in a time when unthinking humans have become their own worst enemy. Civil wars, extreme weather events, degraded water supplies and food insecurity are the outcomes of human actions that can only be resolved by wiser action from individuals using all of their minds. Since this asks for major shifts in both thinking and practice, it requires individuals who can think independently from the mainstream. Individuals who are coping with drastic changes in their own lives are being asked to think of ways to find a way to a just and sustainable future. To make things even more difficult, independent and outspoken individuals are routinely bypassed by the rest of society. The biblical proverb, 'a prophet is not without honour, save in his own country' tells the tale. In the Local Sustainability Project change programmes, the role of the prophet was taken by independent individuals whose country is the planet and who draw on all the dimensions of their minds.

The chapter that follows is a change in direction. David Waltner-Toews explores another side of the human–environment relationship (Chapter 4). It is crucial for our common future that the strong individuals I have been discussing up until now direct their strengths towards working together and to caring for each other. We are all social animals, bound to one another from birth to death; from our mammalian capacity to rear our young to the widespread dread of dying alone. What I may have overlooked in my enthusiasm for individual thinking is the reality of our citizenship within a complex system. David Walter-Toews and Ellen Wall place the individual in a socially oriented perspective.

For the rest of Part I, our story moves on to how strong individuals have worked together to enhance one another's experience of this uncertain world. The chapters in Part II are made up of the work of a group of independent thinkers who shared a common aim for a just and sustainable future for the planet. Before that, John Harris describes the writing group that brought these authors together (Chapter 5). Then we explore the interrelationships between members of the writing group as they share their work over four years (Chapter 6). I think you will find that their combined work tells a story of the ways in which they draw on the multiple dimensions of their own minds, to the benefit of other individuals and groups, humanity and the planet.

Bibliography

Alexander, C., Ishikawa, S., Silverstein, M., Jacobson, M., Fiksdahl-King, I. and Angel, S. (1977) *A Pattern Language: Towns, Buildings and Constructions*. New York: Oxford University Press.

Ashby, W. R. (1954) *Design for a Brain: The Origin of Adaptive Behaviour*. New York: Wiley.

Ashby, W. R. (1956) *Introduction to Cybernetics*. London: Chapman.

Bateson, G. (1958) *Naven: A Survey of the Problems suggested by a Composite Picture of the Culture of a New Guinea Tribe*

drawn from Three Points of View (2nd edn). Stanford, CA: Stanford University Press.
Bateson, G. (1973) *Steps to an Ecology of Mind: Collected Essays in Anthropology, Psychiatry, Evolution and Epistemology*. St Albans: Paladin.
Berlin, I. (1998) *The Proper Study of Mankind: An Anthology of Essays*. London: Pimlico.
Buber, M. (1958/2000) *I and Thou*, Ronald Gregor Smith (tr.) (2nd edn). New York: Scribner Classics.
Camus, A. (1942/1989) *The Stranger*, M. Ward (tr.). London: Vintage International.
Clark, R. W. (1973) *Einstein: The Life and Times*. London: Hodder & Stoughton.
Descartes, R. (1637/1946) *A Discourse on the Method of Rightly Conducting One's Reason*, J. Veitch (tr.). London: Everyman's Library 570.
Doidge, N. (2007) *The Brain That Changes Itself: Stories of Personal Triumph from the Frontiers of Brain Science*. Melbourne: Scribe.
Funtowicz, S. O. and Ravetz, J. R. (1993) 'Science for the post-normal age', *Futures*, 25(7), pp. 739–755.
Heidegger, M. (1970) *Discourse on Thinking*. New York: HarperCollins.
Keats, J. (2010) *Poems of John Keats: Popular Penguins*. London: Penguin.
Lovelock, J. (1979) *Gaia: A New Look at Life on Earth*. Oxford: Oxford University Press.
Marshall McLuhan, M. (1951) *The Mechanical Bride: Folklore of Industrial Man*. New York: Vanguard Press.
Marshall McLuhan, M. and Fiore, Q. (1967) *The Medium is the Massage: An Inventory of Effects*. Harmondsworth: Penguin.
Macy Conferences summary at http://www.asc-cybernetics.org/foundations/ history/MacySummary.htm [accessed 11/2/18].
Mead, M. (1928) *Coming of Age in Samoa: A Psychological Study of Primitive Youth for Western Society*. New York: William Morrow & Company.

Mead, M. (1978) *Culture and Commitment: The New Relationships between the Generations in the 1970s*. New York: Columbia University Press.

Medawar, P. B. (1967/1915) *The Art of the Soluble*. London: Methuen.

Rittle, H. and Webber, M. (1973) 'Dilemmas in a general theory of planning', *Policy Sciences*, 4, pp. 155–169.

Sartre, J.-P. (1946/48) *Existentialism and Humanism*, P. Mairet (tr.). London: Methuen.

Teilhard de Chardin, P. (1966) *Man's Place in Nature*, B. Wall and R. Hague (trs). London: Collins.

Toffler, A. (1970) *Future Shock*. New York: Random House.

Wright Mills, C. (1959) *The Sociological Imagination*. Harmondsworth: Penguin.

4

Celebrating difference

On not losing one's mind

David Waltner-Toews and Ellen Wall

> None lives for himself alone, nor yet merely for others; all live for a community which is theirs. L. Haworth, *The Good City*
>
> Only when difference has its home, when the need for belonging in all its murderous intensity has been assuaged, can our common identity begin to find its voice. M. Ignatieff, *The Needs of Strangers*

In the last decade of the 20th century, our university-based research team received a grant to define, in both theory and practice, what we then called 'agroecosystem health'. Like some of the research networks described in other chapters of this book, our team was initially composed of academic scholars from multiple disciplines. Only later, when we crashed into several intellectual and practical dead ends in both Canada and abroad, did we check our peripheral vision and reassess what we had been doing.

One of the most problematic areas we uncovered in our reassessment was the notion of diversity, initially introduced by team members as a biological phenomenon, but later applied to cultural and social problems. Even within the research team, describing and assessing the agricultural communities with whom we worked required us to acknowledge multiple legitimate, and often apparently irreconcilable, perspectives. At the same time, to characterize and assess whole systems, and working in a context where our government was being called on to make urgent decisions on agriculture-related issues such as tobacco farming, mad cow disease, food safety and pesticides, we found a need to expand the peer group, which had the effect of multiplying and amplifying our differences.

In a general sense, we realized that the necessity to accommodate many diverse perspectives well beyond academia was a foundational assumption of our research agenda. If our survival does not depend on some larger reality, a larger narrative within which our individual mortality is couched, then each of us might as well go her own way, knowing that, when the bell tolls, it tolls for who- or whatever, but not necessarily for us. We can turn out the lights when we leave, since the lights are for us alone. If this were the case, then one could be sorely tempted, on the one extreme, to imagine a 'multi-verse', in which all diverse perspectives are legitimized, regardless of context.

On the other extreme, if we already know the 'right' narrative, then the temptation is to espouse a new global eco-messianism, environmentalism with a tyrannous face. Indeed, there are many in the field of ecological management who, in their zeal to save the planet, gravitate in this direction.

In our readings and discussions with Silvio Funtowicz, Jerome Ravetz and others within a loosely defined international network of scholars dealing with similar issues, we began to understand that we were not alone in grappling with these challenges. Indeed, what was required, we discovered, was a different way to frame the questions, a post-normal science, to address situations in which the facts are uncertain, values are in dispute, the decision-making stakes are high and there is a sense of urgency that decisions be made (Funtowicz and Ravetz, 1993).

Challenges in a multicultural world

When a plurality of perspectives is viewed as a mutual accommodation within a sense of global community or within a liberal-democratic technocratic framework (which is where many of these ideas have been given their modern voice), then the task of doing post-normal science is difficult, but essentially reasonable and civilized. What has become apparent in the past century, however, is that, in the world in which we live, the world of power and culture, of injustice and imperialism, each of those perspectives may protect its own vision of reality with 'murderous intensity'. Thus the greatest challenge in our learning journey to the future may not be to find a technical solution to the energy or water or food problems, but to find a home for difference, even as we find a voice for a common destiny, to weave an unknown collective narrative from our stories, without losing the particularities of our individual minds.

Globally, there are strong homogenizing forces cutting across established divisions of class and ethnic background. The rise of transnational corporations and global markets, drawing on notions of an open-ended, evolving future, have supported a mass culture largely driven by North American values and, until recently at least, effectively undermined the power of nation states to determine conditions for their resident populations. Individual humans have been decoupled from traditional values, tastes and attachments to become what Ignatieff (1984) calls abstract humanity or abstract individuals. As well, the redrawing of national boundaries in the post-colonial age, economic restructuring, war, famine and oppression have uprooted people and resulted in massive movements of populations around the globe.

These population displacements continue to result in de facto ethnically plural countries and, increasingly, plural local communities. It can be a highly alienating process, with one kind of defiant response being the glorification of whatever will confirm people's tribal differences from each other, usually rooted in place-and-culture-specific ideals. We are witness to waves of ethnic nationalism expressed as ethnic cleansing, civil war and racial hatred, some of which results in more population movements and the re-creation of ethnically plural societies – although not necessarily ethnically plural local communities – elsewhere.

These events emerge in the context of an ironic paradox. On the one hand, we (especially those members of the privileged economic minority) are presented with unprecedented opportunities to further personal growth and experience within a global shopping mall of cultural values. On the other hand, we find limits to the satisfaction those choices offer because we lack a meaningful frame of reference that can inform us that we have made the 'right' choice (Ignatieff, 1984). As Canadian poet and singer Leonard Cohen aptly puts it in his poem/song 'Closing Time', it 'looks like freedom but it feels like death' (Cohen, 1992).

In response to this looming wasteland, new social movements have emerged and draw on a well of nostalgia, where the idea of community serves as a kind of social, economic, health and religious icon. Other responses to globalization, not necessarily reliant on ideas of community, claim that a solution lies at the global level, where attempts have been made to establish and enforce universal standards in human rights and environmental integrity or health (Laxer, 1995; Rapport, 1998; Roniger, 1995; Westra, 1994).

Teilhard de Chardin (1959) proposed a more generalized, evolving narrative in which humans have moved from the biosphere to the noosphere (literally, an envelope of thought that surrounds the globe) and that, through the realm of culture and ideas, we can find the potential spaces for a global convergence of human thought and meaningful action.

There is the danger, of course, that such a grand narrative can be usurped by those who believe that this process requires some guidance from an economic and political elite, those who have the 'correct' (i.e. corporate capitalist, Western scientific, Muslim, Christian, etc.) understanding of what is really happening. This usurpation can take place at any scale: in the household, village, neighbourhood, nation, culture.

Kerry Arabena, in her earlier work and in this volume (Chapter 7), has proposed uniquely helpful, alternative ways in which de Chardin's ideas could be made operational, finding a home for difference in the struggle for a globally unfolding narrative. Her work suggests that a locally rooted, global narrative could emerge from mutually respectful engagement between indigenous and scientific knowledge (Arabena, 2015). At the time we were doing our work, however, indigenous perspectives still remained largely veiled from mainstream scholarly culture.

Diversity and the Greater Good

In general, we consider diversity to be good; however, when viewed within a nested hierarchical view of reality, complications quickly become apparent.

As Lister has articulated (2008), the value-laden concept of diversity itself holds the key to understanding just how it can be maintained and effectively incorporated into our understanding of how independent minds can collaborate to generate a global vision.

As humans, we can observe the world as separate from us and note how the variability of individuals is subsumed at the level of species, seeing both the forest and the trees, but at different times and for different purposes.

Implicit in this instrumental concept of diversity is the recognition of boundaries, of limitations (either physical or ideal), that allow us both to recognize differences and to give them meaning. For instance, if we are interested in the diversity of oak trees, the first step in our investigation will be to set limits by including only those trees in the Quercus genus. The differences among red and white oaks are only significant in light of their all being oak trees. If we had decided to consider diversity among all deciduous trees instead, then features that distinguish all oaks from all maple, birch and ash etc. would command our attention and the idiosyncrasies of white as opposed to red oaks would be insignificant. Recognizing diversity implicitly requires a bounded context.

Similarly, with humans, differences of ethnicity and class are only recognizable when individual humans are grouped together in units where the variation has some meaning. Maori, Mestizos, French and Germans are only 'different' when juxtaposed as types with a range of human cultures. The notion of ethnicity becomes important when societies have a recognized form that moves from being relatively homogeneous to heterogeneous. For instance, during most of the 20th century, Anglo-Canadians thought of themselves as 'Canadians' and all others as 'ethnic' or hyphenated Canadians. This perception changed with the existence of growing numbers of non-Anglophones who can also claim Canadian status. Prior to the 1970s, many were marginalized in Canadian society, which was defined primarily by its British elites (Porter, 1965).

In part because of this value-based, bounded context, diversity is a conditional rather than an absolute good. In some scholarly cultures, negative perspectives of diversity prevail.

Robert Putnam (2007) reported that social capital (a term representing the strength of social relationships) was lower among ethnically heterogeneous compared to homogenous groups in the United States. Using measures of trust, altruism and community cooperation, he found that diversity seemed to generate anomie or social isolation. 'People living in ethnically diverse settings appear to "hunker down", that is pull in like a turtle' (Putnam, 2007, p. 149). In other words, it was not just that individuals resisted reaching across and forming ties with those who are dissimilar, but they also lost interest in participating with those who were the same. They opted out of community life in greater numbers than people in more culturally homogeneous settings.

From medical studies, we know that mixing human, animal or plant populations that have not co-evolved in the same geographic area can have devastating impacts on the health of the individuals and populations involved. Evidence for this comes from sources as varied as the historical record (McNeill, 1976; Wright, 1992) and from epidemiological studies of emergent infectious diseases (Lederberg et al., 1992). This is not surprising, since key ingredients for an epidemic include both exposure to a potentially pathogenic agent (of which there are a great many) and a pool of susceptible, previously unexposed hosts (Fine, 1993). Furthermore, we have all been lectured about the negative consequences of invasive species, which, in some cases, may increase diversity.

Because of the context dependence of the impacts of diversity, there is a real danger in looking at studies of specific institutions, places and times (20th-century United States, colonial empires, corporate, large-scale agriculture) and from them to claim general truths, to confuse what *was* or *is* in specific situations with what *could* or *should* be.

Depending on the sorts of institution and community we create, we can use diversity to nurture the best, or promote the worst, of human experience. The nature and structure of the communities within which individuals are mixing and the types of social support network available to them influence the probability of epidemics, buffer the negative effects from illness and other kinds of stressor and create conditions for a more resilient society (Cassel, 1976; Fine, 1993). Thus, on the one hand, diversity may lead to epidemics and, on the other, to resilient public health programmes and sustainable communities.

The cementing of difference can create vile institutions such as slavery and apartheid. But the resolution of difference can, as well, lead to human achievements of the highest order. Work completed by the Truth and Reconciliation Commissions in South Africa is a good example, along with practices of narrative therapy and just therapy in New Zealand and Australia. Practitioners of the latter have shown, through careful, mutual listening to narratives across cultures and power groups, the possibilities of a more convivial social-ecological culture (Waldegrave et al., 2003; White, 2003). Both abuser and abused members of closed families and communities can lose the worst excesses of patriarchy, incest, dogmatism and other forms of abuse, but be able to carry forward the best of their traditions – a sense of place, the common good and self-worth – into a new, more inclusive, sustainable community.

In Canada, various social movements of the 1960s, multicultural policies and ideology and the Charter of Rights and Freedoms have ensured, at least in theory, that the context has broadened enabling the country to become a more explicitly diverse society. The move to recognize and give legitimacy to Canadian diversity by institutionalizing multiculturalism into an official policy in 1971 and then into an Act of Parliament in 1988 occurred in the same time period as the strengthening of the Canadian state through a repatriated constitution and Charter of Rights and Freedoms. These events and developments made the implicit context for diversity into an explicit, codified reality. They were not coincident by chance but out of necessity, to keep the potential for division in check. In the context of the tension or contradiction that exists between the community as a whole and its constituent individuals, between the constraints set by the needs of the collective and the freedom/autonomy people need to be fully human, state guarantees for equal rights and multiculturalism are fundamentally dependent on one another.

The Canadian example illustrates the continual interplay between diversity at one scale (individuals) being contained within, and being necessary for, unified classes at larger scales (family groups), which themselves constitute diversity at another scale (community). The tendency of those larger units is necessarily to dampen internal disorder or chaos that may accompany the urges to diversify (Bella, 1994).

Nurturing a creative, global mindset will thus be best served by having a great many different, independent ways of thinking and to create situations in which new organizations emerge to contain the diversity and give it meaning; in this way, creative sources for future adaptation are protected and available for future needs.

Where do we go from here?

Because human–environment interactions take place in specific places, with particular plants, animals, soils and other resources, rather than at some abstract global level, communities of place are also essential to foster the biophysical health of an eco-social system. This is at the root of Jane Jacobs' ideas of 'the nature of economies' (2000). Biodiversity can only be defined in terms of both place (for instance a variety of plants and animals in a watershed at any given time) and time related to that place (such as crop rotations, seasonal or longer term changes in biota and the changing needs of residents). The human communities that exist in ecosystems serve as a collective, self-conscious memory and comprise the cultural equivalent of genetic information. It may be that both the land and the individuals living on it will continue to suffer as people's attachments to place are exchanged for linkages to more superficial global cultures and transient communities of interest (Berry, 1977).

As we have suggested, diversity of thought – as much as biological diversity – brings with it serious challenges for system sustainability. Effective responses to those challenges rely on incorporating scale and context into our understanding, while recognizing that diversity is inextricably linked to notions of unity. Our future, as Arabena suggests, lies in reconnecting people to place and nurturing their sense of self through the authentic experience of belonging to, and caring for, something material and permanent. At the same time, any 21st-century revival of communities of place, to make positive contributions to human well-being, must be significantly different from those of the past, when ethnicity, culture and territory, defended with murderous intensity, formed the basis for racial imperialism. Instead, the communities of place we need are those where diverse communities of interest, emerging from different cultural histories, co-exist and enrich each other in multiple and emerging forms, recognizing their fundamental dependence on one another.

A sustainable future for humanity depends on how individuals define themselves as persons and experience a sense of dignity. Rather than looking inward and sensing our intrinsic value by asserting our own, indivisible rights, we would be better served by reconceiving the ideal, so that personal dignity is realized through our association with people and things beyond individual selves, embodied in our community of place. Collectively, it is incumbent on us to create systems and institutions that will give legitimacy to individuals' 'self-transcendence'. Referring to such a process as the reconstruction of social identities without eliminating their personal importance, Putnam (2007) points out the adaptation involved is not just between different cultural groups, but to the new society that inevitably emerges.

The tensions between independent thinkers and the communities of which they are a part will never be – and probably should never be – resolved. A sustainable, healthy human community will be one in which we act on the faith that there is a common narrative, a common global journey, but that this narrative is not predetermined and only becomes visible as we tell it and live it. The human story – our planetary story – is an emergent tapestry of tales told from many perspectives, rooted in a common reality whose nature and dimensions we can perhaps, in solidarity, collectively – and only collectively – discern.

In Chapters 11 and 17, I describe initiatives that grew out of the lessons we learned from our agroecosystem health reassessments. Those initiatives, as well as the many other case studies reported in this book, demonstrate that this engaging of individual thinking to create an emergent, common good is a challenging, but not quixotic, task.

Indeed, in the following chapter, John Harris describes one way in which the tensions among independent thinkers was navigated through having them learn together and in so doing create a community in which their differences are not only acknowledged, but strengthened.

Acknowledgements

The authors of this chapter wish to acknowledge the members of the multidisciplinary Agroecosystem Health Research team, and the financial support of Canada's Tri-council Eco-research program.

References

Arabena, K. (2015). *Becoming Indigenous to the Universe: reflections on living systems, indigeneity and citizenship.* Melbourne: Australian Scholarly Publishing Pty Ltd.

Bella, D. A. (1994) 'Organizational systems and the burden of proof'. Paper presented at the symposium: Pacific Salmon and their ecosystems: status and future options. University of Washington, Seattle.

Berry, W. (1977) *The Unsettling of America: culture & agriculture*. San Francisco: Sierra Club.

Cassel, J. (1976) 'The contribution of social environment to host resistance'. *Am. J. Epidemiol.*, 104(2), pp. 107–123.

Cohen, L. (1992) www.leonardcohensite.com/songs/closing.htm

de Chardin, T. (1959) *The Phenomenon of Man*. New York: Harper & Row.

Evans, R. G., Barer, M. L. and Marmor, T. R. (1994) *Why Are Some People Healthy and Others Not?* New York: Aldine de Gruyter.

Fine, P. E. M. (1993) 'Herd immunity: history, theory, practice'. *Epidemiologic Reviews*, 15(2), pp. 265–302.

Funtowicz, S. O. and Ravetz, J. R. (1993) 'Science for the post-normal age'. *Futures*, 25, pp. 739–755.

Haworth, L. (1963) *The Good City*. Bloomington: Indiana University Press.

Ignatieff, M. (1984) *The Needs of Strangers*. New York: Penguin Books.

Jacobs, J. (2000) *The Nature of Economies*. Toronto: Random House.

Lacy, W. (1994) 'Biodiversity, cultural diversity, and food equity'. *Agriculture and Human Values*, Winter, pp. 3–9.

Laxer, G. (1995) 'Social solidarity, democracy and global capitalism'. *Canadian Review of Sociology and Anthropology*, 32(3), pp. 287–315.

Lederberg, J., Shope, R. E. and Oaks, S. C. (eds.) (1992) *Emerging Infections: microbial threats to health in the United States*. Washington, DC: National Academy Press.

Lister, N.-M. (2008) 'Bridging scales and values: the challenge of biodiversity conservation', in Waltner-Toews, D., Kay, J, Lister, N-M. (eds.), *The Ecosystem Approach: complexity, uncertainty, and managing for sustainability*. New York: Columbia University Press, pp. 83–107.

Maffii, L. (2006) 'Bio-cultural diversity for endogenous development: lessons from research, policy, and on-the-ground experiences'. Paper presented at Endogenous Development and Bio-Cultural Diversity. Geneva Switzerland, October 3–5. Retrieved 10 November 2008, www.bioculturaldiversity.net/Downloads/Papers%20participants/Maffi.pdf

McNeill, W. H. (1976) *Plagues and Peoples*. Garden City, NY: Anchor Books.

Porter, J. (1965) *The Vertical Mosaic: an analysis of social class and power in Canada*. Toronto: University of Toronto Press.

Putnam, R. (2007) 'E pluribus unum: diversity and community in the twenty-first century', the 2006 Johan Skytte Prize Lecture. *Scandinavian Political Studies*, 30(2), pp. 137–174.

Rapport, D. (1998) 'Ecosystem health: an integrative science', in Rapport, D., Costanza, R., Epstein, P. R., Gaudet, C. and Levins, R. (eds.), *Ecosystem Health*. Oxford: Blackwell Science, pp. 3–50.

Roniger, L. (1995) 'Public life and globalization as cultural vision'. *Canadian Review of Sociology and Anthropology*, 32(3), pp. 259–285.

Waldegrave, C., Tamasese, K., Tuhaka, F. and Campbell, W. (2003) *Just Therapy – a journey*. Adelaide: Dulwich Centre Publications.

Waltner-Toews, D., Kay, J. and Lister, N-M. (eds.) (2008) *The Ecosystem Approach: complexity, uncertainty, and managing for sustainability*. New York: Columbia University Press.

Westra, L. (1994) *The Principle of Integrity: an environmental proposal for ethics*. Lanham, MA: Rowman & Littlefield Publishers, Inc.

White, M. (2003) 'Narrative practice and community assignments'. *International Journal of Narrative Therapy and Community Work*, 2, p. 1.

Wright, R. (1992) *Stolen Continents: the Americas through Indian eyes since 1942*. Boston, MA: Houghton-Mifflin.

5

Multiple minds

The more we are together

John A. Harris

In this chapter, I explore the experiences of members of a collaborative group, the Fenner Writing Group. One question to answer is the extent to which participants surrender their autonomy to the group. Another is whether it was necessary to achieve consensus in order to work together. In collaborative groups, it is usual to agree on a discipline, multiple disciplines or a transdisciplinary approach. This chapter contains an account of a collaborative writing group formed to encourage members to help each other pursue their separate interests. Their work makes up Part II of this book. I write this chapter as one member of this group. The underlying belief in forming the group was that the stronger the individual in her own right, the stronger the group. That is, independent thinking makes a group stronger. To help me write this chapter, members of the collaborative group were asked two questions: 1) What have been your personal experiences in the group; and 2) What was your experience of the group as a whole?

The group was formed in early 2014 when the lead editor of the present book thought that the time was right to live a decades' old dream. After supervising dozens of PhD scholars over many decades, she wanted to run her own race as a writer in a team with other independent writers. The prompt was a small group of her successful PhD scholars wanting to write a book for a general audience that was based on their theses. They were invited to form a collaborative group in the knowledge that they all had an existing background in writing and could help one another with constructive comments and suggestions. After the invitation was greeted with enthusiasm, she entreated the group's then facilitator to 'help me shut up and not allow others to treat me as a supervisor'.

After being invited to join, I was eager to improve my writing. I also wondered whether I was capable of thinking independently and as a member of the group at the same time? I saw this as a paradox and hoped that I could overcome it. I knew that I could access my own biophysical, socioeconomic, ethical, artistic and sympathetic ways of understanding. They would be there for members, too, but I wondered whether the multidimensional mind would be revealed in our personal reflections.

The Fenner Writing Group met for the first time on 31 March 2014 under the auspices of the former Director of the Fenner School of Environment and Society, Australian National University, Professor Stephen Dovers. At this first meeting, all members agreed that the group would be self-organizing. All decisions about how the group was to run and what it could do were in the hands of members. There was collective agreement that group membership was based on the following principles:

- keeping the group small to enable us to read and give space for one another's work
- accepting all forms of inquiry
- creating a safe space to develop a community of learners and fellow collaborators
- valuing a diversity of perspectives.

In addition to acknowledging these principles, the group agreed to co-learning, where, in their formal work and group processes, members help one another to learn. Everyone was writing major non-fiction texts, in the form of a book, PhD thesis or blog. It was agreed that one or two members would contribute work for review at each meeting and circulate it by email before the meeting. Two members, who were experienced and skilled facilitators, were then asked to facilitate our meetings.

After being invited to join the group, one member wrote: 'It's very exposing to put your work up for scrutiny but, at the same time, I loved the idea of a group of peers gathering together to share ideas and provide constructive support.' Another member felt hesitant, too: 'I initially was a little intimidated by the intellectual power of the group but the humanity within the group overrode any misgivings I had.' The reciprocity and shared sense of responsibility shown here led to members becoming acutely receptive and responsive to the needs and aspirations of one another.

In the group process, all comments were directed towards meeting the presenter's goals for their work. This allowed members to be heavily critical of each other's work, if necessary, on the understanding that they were supporting the aims of the writer. To ensure this, each meeting started with the presenting author being asked what they hoped to receive from the discussion. Each reviewer was asked to start with a positive response, before proceeding to the critical evaluation of the work. In this way, a principle of critical loyalty was established by agreement to support one another's work through constructive criticism. A member wrote: 'I was immediately encouraged at the first meeting I attended by the warmth and genuineness of the group and the ethos of critical loyalty.' Another member told how this 'led to robust and honest feedback and more helpful outcomes for the writers'.

At the same time, each member was accorded equal respect, rejecting any trace of a hierarchy built on experience, age, qualifications, previous recognition or gender. It was expected that every member could trust every other member, with confidentiality for all group discussions and security for creative ideas. A member wrote:

> Our two facilitators have been integral to the flow of our meetings and guided us through the shoals of idle chatter to the heart of what the participants sought from discussion. Their styles of facilitation were different; however the effect of their light-handed facilitation was to engender a feeling of trust.

Reviews of one another's work could be made beforehand on a document circulated by the author to the whole group and this formed the basis for appreciative and critical discussion at meetings. One member reported:

> Each time we met the discussion would delve deep into issues and this can only happen when there is trust so members feel comfortable to reveal sometimes deeply personal aspects of their writing or experiences of life.

To my mind, this was the source of the ideas expressed in Part II of this book. It was early 2018 when this idea of members contributing chapters to the book emerged. For these chapters, as well as for the general exchange of critiques, members agreed to become external reviewers of one another's work being submitted for the book. At these times, formal notes were taken of members' comments and recommendations. These were circulated to the authors and the whole group for accuracy and follow-up.

In addition, it was agreed that authors' acknowledgements should be both academic and personal. The published work of members already included two PhD theses, three books, numerous journal articles and a very popular blog.

Monthly meetings were held at the local botanical gardens, looking out on a green canopy. Occasionally, a member offered their home. Meetings began with members sharing what had happened in their work and lives since the last meeting. Reflections could include significant personal insights and achievements with members free to choose what to share. Meetings generally ran for three to four hours with a break for lunch. They began mid-morning to enable members who lived outside Canberra ample time to get to the meeting safely. Over the four years, the group changed as members moved away and new members were invited to join. There was always a core group of five to six members who met monthly; others came and went over the four years. The aim was for there to be no more than nine or ten in the group at any one time, giving time for discussion, and not too heavy a workload. A member wrote:

> I have loved being part of the evolution of our group, how each of us plays to our unique style as it jives with the spirit of our group, as it develops its own rhythm and beat. Our group's field is a stable core, and generously flows with comings and goings.

Building a unique container

The writing group created a container in the sense of Isaacs who refers to a field as 'the quality of shared meaning and energy that can emerge among a group' (Isaacs, 1999, p. 242). He recognizes four key aspects of a container as the active experience of people listening, respecting one another, suspending their judgement and speaking their own voice. The member's comments reported in their individual reflections support that this was true for the writing group. Isaacs is a strong advocate of Bohmian dialogue (Bohm, 1996) and the founder of the Dialogue Project at the Massachusetts Institute of Technology. Dialogue comes from the Greek word *dialogos*, where *logos* means 'the word' and *dia* 'through'. The image arising from this derivation is of a stream of meaning flowing among and through us as participants think together. Bohm believed that it is this flow of shared meaning, unspoken, that holds a group together and provides the fertile ground from which individual and collective thought emerges. Bohm's genius was to realize this and that multiple points of view must be expressed. This opens new possibilities and multiple options that he called *generative dialogue*.

The writing group did not generate its own container at first. It required skilful facilitation and the participation of all members before one could be built from the flow of the meeting and the autonomy of each of its members. In essence, there were two containers built. There was the collective container of the group itself and the individual as a container for their own developing work. What we all brought included unspoken and formal knowledge from growing up in Australia with its aim for democratic governance. There were also our innately human characteristics, including capacity for learning and our multidimensional minds. As the group continued, other attributes emerged, including collaborative learning, independent thinking and openness to change.

Once these collective and individual containers were established, it was not long before the group became self-organizing, allowing the creative processes of the group to emerge. I had experienced this occurring from my previous academic experience when I participated in the design of a new undergraduate course in environmental science. In this course, the best expression of student group work occurred on field study trips, often in remote locations in the Australian bush. I realized two important things from my experience on field trips that are relevant here: first, that independent thinking can flourish in a collaborative group; second, it was independent-minded individuals who acted as catalysts for change. Now, as a member of the writing group, I participated in building a container for individual autonomy and independent thinking. One member commented:

> I have been amazed that almost all writers seem to have moments of anxiety about the worthiness of their message or even whether they have the right to 'put things out there'. I have certainly gained some confidence in my own writing since being involved in the group and have often felt as though I have been some help to others when I have made comments.

We were fortunate to have two highly skilled facilitators as members. Each had their own professional and personal styles. Julia's facilitation approach was sourced from process work and deep democracy – 'a philosophy, an attitude, and a facilitation method, pioneered by Arnold Mindell and colleagues around the world' (Mindell, 2002; 2014). She explained how she sources information from three interconnected levels of experience: objective, subjective and atmospheric:

> For the writing group, there was the measurable objective consensus domain that had to do with theses, books, blogs and manuscripts; publishers, meetings, time frames, deadlines, feedback and so on. The subjective experience was generated from individual group members by the ways each expressed

> themselves and met together each month. This could include a polarized perspective on a topic, a moment of joy; little flare-ups, irritations, as well as inner and outer critics and one's allies in writing and reflecting on our own writing. The atmospheric level is a vaguer, hard to describe essence level of experience.

Julia explained how the essence level 'is present in the transformative effect of a piece of writing, a bond we may feel for each other in a given moment, and a feeling of connection with the larger evolving story of humanity'.

The other experienced facilitator of our group, Craig, kept continuity of communication between members over the four years. He described his style of facilitation as 'catalytic', because he saw himself as totally dependent on the input, values and effort of others. He also took on the role of 'instigator' and 'being willing to support risky, subversive or transgressive thinking'.

Craig chaired most meetings and his easy-going style and openness of facilitation complemented Julia's in many ways. So much so that when they attended a meeting together, she told how they would: 'strike a similar note, a kind of improvised jazz, playing off each other noticing what the other is noticing and checking with each other in the midst of the action'. From Julia's viewpoint: 'A deep democracy attitude, such as the writing group, reflects the diversity of human experience, with all its tensions, contradictions, and mystery.' This viewpoint aroused my curiosity about how best to change the way I was participating in the writing group.

The first year with the group was a very significant learning period for me when I opened my mind to a new way of being in a collaborative relationship with others. I realized for the first time that the relational field of energy between us was a boundless 'I–thou' universal relationship and not an abstracted 'I–it' relationship of thinking of others as separate entities, disconnected from the whole (Buber, 1958). I saw the group as made up of independently minded individuals and at the same time a collective whole. I had a sense of what was at stake for me in the group. I had this wonderful feeling that here was an opportunity to learn anew and change habits that I had built up over a lifetime, which included 50 years of studying and working as a scientist and academic.

Feeling meaning is central to skilful practice (Walkerden, 2005). For this to happen, one has to pay close attention to how they are personally experiencing a situation. I found in my work as a scientist that my own experience was largely outside the bounds of professional discussion and that conversation about such matters was considered personal. It was clear to me in my professional life that how I felt about my work was very important to me and that individual thinking goes on for the most part without the use of symbols or words. Albert Einstein thought that 'wondering' about something is feeling and the place where fresh insight comes from (Einstein, 1949/2002). Similarly, when jazz musicians improvise together they 'feel' where the music is going.

In early meetings I felt a strong need to speak my mind because I had seen my role as helping others based on my extensive teaching experience as a university academic. I had reverted to this role, not realizing that it was detrimental to this new kind of relationship that included my own learning. I would come up with what I thought was a good idea and want to share it; otherwise, I became impatient and anxious. It was only after opening my mind and listening more attentively to what was happening in the group that I became aware that I must make the change to feeling a new situation happening in the group, and to learn from the others.

I required help to change. I did not realize what I was doing until it was pointed out to me. I had been unconsciously defending my own ideas, opinions and prejudices as though I knew best. This had led to my frustrations. What had once been meaningful for me in my teaching role had become fixed and incoherent in the new situation. My defensiveness also affected the flow of meaning in the group, which required everyone sharing their different points of view freely.

From Julia, I learned that tension and friction in the group is 'flow that is interrupted and trapped in polarization'. She explained:

> In our writers' group, we were learning to welcome tension, make visible the polarization in the moment by stopping, asking and giving space to unfold the interaction between different sides of the issue, be that an idea, a momentary relationship tension or, maybe, a deeper principle that someone feels strongly, and can't budge from. All are doorways into something new.

This understanding helped me accept conflict as positively reflecting our individual differences and inherent complexity. I learnt not only to follow the flow of the group but also think about what it meant for me and the group as a whole. I could contribute in a different way and overcome the paradox that I saw when first joining the group of 'thinking independently and as a member of the group at the same time'.

In another example of deep learning, we met at a member's home and were welcomed with morning coffee. There was also lunch during a break in proceedings. A member of the group had presented a personal piece on the difficulties of writing and had asked members for their critiques. An hour later, there was tension and friction in the air because the author had observed that nobody had given her constructive criticism. Others disagreed. With facilitation, the flow of discussion returned and everyone told how they respected the author so much they would not criticize her writing. This, in turn, caused acute frustration. She believed that in order for her to keep learning and improving her writing, it was crucial for the work to be critically evaluated. The meaning of critical loyalty became coherent in that meeting and the unspoken tacit ground of the group from then on.

If I were to be a better participant I had to listen, feel the situation and speak less. I needed to keep my thoughts that arose from past experiences separate from my thinking in the here and now; to face up to the challenge of staying in the moment and open to all levels of experience. I knew that I could put my trust in my intuition for grasping the flow of meaning provided I was attentive. There were also occasions in the writing group when, suddenly, I had an insight and my understanding had taken a leap forward. Nevertheless, I still had to remain patient and contribute in my turn. Through personal reflection afterwards, I discerned that these insights were always holistic and multidimensional.

Bohm advocated dialogue for reuniting the individual, collective and cosmic connections of human beings. He believed that for most of human evolution multiple ways of thinking had prevailed. With the scientific era giving emphasis to thinking objectively there was a devaluing of the subjective experiences of individuals and other ways of thinking. This helped to generate the fragmentation of thought current in the world today. It is well known that humans rely on language, acts, gestures and interpretation of context to understand the minds of others (Bruner, 1996). If dialogue could become generative,

Bohm foresaw a more coherent form of thought emerging in the world. It would give people a sense that boundaries are permeable, that objects have an underlying relationship and that there is a participatory quality to existence. With respect to our group, provided members held to their independent thinking, our viewpoints became part of a collective flow of meaning and coherent. Like a stream, the flow moves between each individual mind, forming a collective mind.

There is no doubt that the writing group offered me a new way of looking at things, of bringing my feelings into words and enabling me to give them new life in my imagination. I gained a strong sense of goodwill for the group members over the four years. I was able to expand my mind, see through their eyes and feel their thoughtfulness. It was exciting and stimulating for me as I was pitched out of my everyday world to experience myself and the others anew. This degree of shared experience also gave me a strong sense of community, which carried over into the wider world. As another member commented: 'I made some very firm friends within the group who I value highly.'

Hanging in there

Over the four years, members found their own personal and distinctive ways of participating in the writing group. Craig gave an example to illustrate what he called the diversity and unity of the members. Referring to a meeting spent reviewing the outline of his PhD thesis he wrote:

> I found the group very encouraging and they dissected the outline with gusto. All were very helpful except there was virtually no agreement. The discussion was full of good humour and friendly disagreement. This gave me a visceral understanding of the problem I faced with my thesis.

Another member agreed:

> For me, the expertise of the individual group members is overridden by the different views of each, which, together, give a multifaceted whole and an invaluable insight into what readers will feel as they read my work. Our two facilitators allowed the space for ideas to come out and the confidence we have to express our views in an atmosphere of acceptance and trust, making the dynamics of the group more powerful than the comments of any one reviewer.

This was put differently by another member, writing:

> The diversity of professional experience and affiliation is not only reflected in the insightful and acute comment on my own work but exposes me to a broad range of work to review. I have learned a lot from reviewing the work of others.

Reflecting on the group as a whole, Craig listed five qualities of the group:

> First, there were strong bonds from the networks that members belonged to, including everyone's love of nature. Second, there were other tacitly held values about people in general and, more specifically, of treating each other as whole multidimensional individuals. These meant members were conscious of how their comments may affect another member. Third, only having two or three members present their work at meetings allowed ample time for questions and extended discussion. Fourth, there was a genuine appreciation of individual differences with the writer left to decide what is best for them to do. Fifth, the atmosphere of the group was relaxed, with room for good humour. Yet, members had the time to think seriously and deeply on the issues raised.

One member raised the question: 'I wonder what the difference was between those who hung in there and those who came and went?' She went on to answer her own question:

> I noticed that the core members made great efforts to attend every meeting, re-orienting their diaries in order to come. They regularly provided comments on each other's work. To me, these members, who hung in there, seemed to develop a shared body of work and a strong camaraderie.
>
> The sporadic attendees seemed to be less interested in the other members' topics. Some gratefully received comments on their own work without giving comments to others. A couple of times it was clear the members had not grasped the idea of critical loyalty. One gave a very unsympathetic critique with no suggestions on how the paper could be improved. Another described how other people's work was of no use to them, not asking whether it met the aims of the author. It was useful to reflect on this, to know things to avoid in my own work.

Thinking for oneself

Thinking for oneself was safeguarded by the protective presence of the collective container. In particular, it supported independent thinking with honest, constructive criticism. This was not always the case in early meetings. As one member commented: 'If you are not honest in your comments it is not helping. Critical loyalty is about receiving honest feedback and not taking it as a personal criticism.' Another member told how she valued feedback that has 'the ferocity of love and a sentiment of wanting to make someone really good'. In addition, she found that the academic focus of some of the writing somewhat restrictive, as she was working in a more applied field.

One of our facilitators believed that the role of facilitator could be taken up by anyone in the group who 'has the awareness in the moment to name the feeling, notice the hesitation, appreciate the tension and catch the process moving us'. This was demonstrated whenever one or more individuals in a group stepped outside the experience of a difficult situation, restoring dignity, respect and the feeling of wholeness to a group.

There was a dramatic example of this happening when the principle of critical loyalty was breached, coming close to being a personal attack. In this case, there was an immediate group response, unspoken, that calmed the situation. At the next meeting, this event was processed and discussed by the group. On another occasion, there was the suggestion that a member was deferring to authority and weakening their independent voice. In contrast, other members held to a deep principle that they felt strongly about and would not budge. In both these cases, facilitation opened up discussion far more than might have been expected if the group had not formed the container.

One perceptive member reflected:

> The dynamics of the group and the personalities within it made it a forum in which we wanted to immerse ourselves. The opportunity for us to have feedback on our writing was a special gift we all cherished. The empathetic atmosphere and the honesty of comments attracted us in to this supportive cocoon from which emerged our writing, like the unfolding wings of a hatching moth.

All together

Meetings always began by going around the group and sharing in turn what had happened since the last meeting. This helped ensure the relationship between members felt right before commencing our more detached and rational discussion. A member told how she liked the way group members updated their life story in this way: 'It creates closeness and an understanding of each other, and is very supportive. It is a great empathy and like-mindedness that contributes to the genuine experience.' Another member wrote how it was possible to 'have a way of feeling into our texts and getting to the essence of what we are trying to reveal to the world and ourselves'. Such feeling was important for strong independent thinking and group participation.

A member reflected:

> Academic writers who, over many years of writing, were confined by the shackles of convention have striven and succeeded in finding their own voice through encouragement. Those who were unsure of the value of their own comments felt more confident when others expressed how their comments were of great value because of their unfettered, clear-eyed view.

Another recalled:

> I always felt included and there have been many positive comments regarding my ability to see somethings clearly from a fresh view outside the academic world. I have been very impressed at the inclusiveness within the group and have felt privileged to be part of it.

I remain puzzled that the multidimensional mind went largely unacknowledged over the four years. Do we take for granted such universal attributes accessed through intuition and insight? My extended education and training did not prepare me well for the acceptance of such inherited gifts. I gave little consideration to 'how do I think?' and 'how can I know?' Usually, I accepted my situation and fitted into what was expected of me according to the circumstances. It is only recently that I have set out to find out more about how to use my mind well and its capacity for multidimensional thinking. Once it was pointed out to me that I had this capacity, it immediately made sense to me as part of everyday life. It was like the joy I found in my findings on the bat cave that formed the basis of my doctoral thesis and the amazing world opened for me.

Having a sense that we are all on the same page was a powerful feeling that I had sometimes as I was participating in the group. On reflection, I realized that it was our individual differences that made it such a rich feeling. It was a paradox. Delving deeper and trying to understand its nature has made me realize that this feeling of inclusiveness is likely because all dimensions of our minds are being expressed freely and openly. We found the biophysical dimension of the garden setting in which we held our meetings to be enriching. The socioeconomic aspect was eating together in the café there; the ethical aspect in the way we valued what we were doing and learning. The artistic ingredient was also there in our humour and our sparking off one another. The sympathetic in how we cared for each other and the wider world of fellow travellers.

Conclusion

The writing group proved to be strongly built on the strength of each member with their different points of view, thus giving support to the belief that the strength of the individual makes for a stronger group. The different personalities, experiences and fields of endeavour proved to be positive synergistic building blocks of our container. To my mind, our discussions and related dialogue stimulated our thinking and the flow of meaning in the group became more coherent. Members were challenged to become more aware and able to think independently. In this way, I learned to be critically loyal to myself, as well as other members and always open to new possibilities.

One member wrote: 'Little did I know what a journey I was about to embark, for me it was a journey of personal and mental growth.' Another member made the point that the group's acceptance of all forms of inquiry meant that her work 'was scrutinized from a range of perspectives and mindsets'. Adding the group 'was an exemplar of the marriage of theory and practice so vital to our work and our time in history'. In Part II of the book, this work is presented by members in their different styles of writing.

Heightened emotions, feelings and relationships flowed back and forth in the group. Meetings became special times when the purpose-built literal thought of society was pushed into the background of our lives. Creativity, intuition, imagination and experience were as important as goal directedness criteria of efficiency and working according to the rules. One member reflected:

> Because of the different views expressed in this group, my feeling is that we have all grown as human beings and also as writers from the whole feeling of the group whose value was always greater than the sum of its parts.

The lead editor concluded:

> The experiment was a great success from my point of view. It took a few meetings and an outburst to feel I had moved. One group member said they had moved 80 per cent. I have found great stimulation and personal satisfaction in being able to receive comments and criticism from my own peers. I have also found it fun and gained existing colleagues as new friends.

There was no doubt that the writing group was a great success provided members hung in there, thought for themselves and all came together as both individuals and a group.

References

Bohm, D. (1996 [2004]) *On Dialogue*. New York: Routledge Classics.

Bruner, J. (1996) *The Culture of Education*. Cambridge, MA: Harvard University Press.

Buber, M. (1958/2000) *I and Thou*, Ronald Gregor Smith (trans.), 2nd edn. New York: Scribner Classics.

Einstein, A. (1949 [2002]) 'All concepts are freely chosen posits', in Steel, D. S. (ed.), *Genius: In Their Own Words: The Intellectual Journeys of Seven Great 20th-century Thinkers*. Chicago: Open Court.

Harris, J. A. and Deane, P. (2005) 'The ethics of social engagement: learning to live and living to learn', in Keen, M., Brown, V. A. and Dyball, R. (eds.), *Social Learning in Environmental Management: Towards a Sustainable Future*. London: Earthscan.

Isaacs, W. (1999) *Dialogue and the Art of Thinking Together*. New York: Currency.

Mindell, A. (2002) *The Deep Democracy Open Forums: How to Transform Organizations into communities: Practical Steps for Resolving conflict and Creating Community*. San Francisco: HarperCollins.

Mindell, A. (2014) *Leader as Martial Artist: Deep Democracy Leadership in Conflict Resolution, Community Building and Organizational Transformation*. San Francisco: Deep Democracy Exchange.

Walkerden, G. (2005) 'Felt knowing: a foundation for local government practice', in Keen, M., Brown, V. A. and Dyball, R. (eds.), *Social Learning in Environmental Management: Towards a Sustainable Future*. London: Routledge.

6

Multiple voices

So say all of us

Valerie A. Brown and John A. Harris

A popular misunderstanding about independent thinking is that it is a rare event, a flash in the pan. In reality, it is the birthright of every human being. I know that I can think. However, in Western society with its admiration for objectivity, it may not be recognized as a regular mode of thinking. Even when thinking outside the square has been recognized, the thinker has usually travelled a familiar path: they have a new idea, it is at first greeted by rejection, and later may find acceptance. However, new ideas are often not recognized at all. They have been blocked by established prejudices and the originators lose confidence in their own thinking.

Another mistaken idea is that there will always be tension between independent and specialized thinkers. The experience of the Fenner Writing Group outlined in Chapter 5 suggests the opposite. The members of the writing group reported that the more confident they became in their writing, the more they felt part of the group. The reverse was also true: the stronger the group became, the stronger the feeling of independence of its members. This double effect is found in the strength of the independent pieces from Fenner Writing Group members in the chapters in Part II.

Further, independence does not necessarily mean rejecting others' thinking. Disagreement between colleagues can be a starting point for collaboration. In one of the case studies in Part I, a team made up of strong personalities with differing interests found it impossible to collaborate. Tensions increased until the project stalled. This forced the team to reconsider the issues through all the multiple dimensions of their minds. Once they were able to share feelings as well as facts, the team worked together harmoniously for a further five years.

Another trap for thinkers who wish to remain independent is to accept the interpretation of a problem from the society that caused it. The way it has been interpreted will almost certainly be part of the problem. The problem will need redefining independently. In a lead-smelting town whose children had a high level of lead in their blood, the suggested solution was for housewives to clean their houses thoroughly. At the same time, their children were playing in soil that had accumulated lead from the smelter for decades. Major influences had reallocated the problem from the owners of the mine to the housewife. Independent thinking would take it back to the smelter.

The Fenner Writing Group's chapters in Part II have allowed us (Valerie and John) to explore the relationships between strong individuals and a strong group. These authors' independent thinking is not a flash in the pan; it permeates each of their chapters. Individual members of the writing group have helped each other go past existing prejudices and release their innovative ideas. Potential for friction had been turned into collective energy. Together, the writers offer ideas for a common future. Their work weaves a new language for our time. Words that loom large in the chapters of this book and yet are unfamiliar to our present society have built a new language for the Anthropocene. The new language is made up from concepts such as post-normal science and post-normal reconciliation, *sophia*, a circular conundrum; collective and independent thinking; organic, mechanical and emergent minds; self-organization of individuals, society and environment; the Western knowledge cultures; a multidimensional mind; an undivided mind; allowing; regeneration of a landscape and; regeneration of society; transdisciplinarity; transcoherence; wicked problems, person centred and many more. These words can be found throughout this book. Although they may need explanation for this generation, they are ready for use in the next.

Here we use this language as we weave a story as told by the Fenner Writing Group, in the order as listed in this book's contents pages. John and I explore how the chapter authors accessed the multiple dimensions of their minds. We know that the writing group members are strong individuals from the chapters that they have written. We know that they are strong supports for one another, as their acknowledgements to the others make clear. Next, we explore their chapters as they tell the story of their combined contributions to an uncertain world.

The story begins in Chapter 7. Kerry Arabena, a Meriam woman and a professor of indigenous health, reminds us that we are all citizens of the same universe. She writes that we need to reconcile the divisions between the components of the community that is the planet Earth. We cannot mediate the present situation as though there were some minimal balance already existing that could be slightly modified – the violence already done is on a scale beyond acceptability.

Kerry recommends that post-normal science be extended to create a post-normal society in which there is equality, friendship and environmental sustainability. We have already quoted the co-founder of post-normal science, Jerome Ravetz, as saying that post-normal science is needed whenever the facts are in dispute, values are contested, knowledge is uncertain, decisions are urgent and decision stakes are high. He could have been describing the conditions in the world of today. Kerry offers the challenge that we all, working together, have the capacity to reconcile our differences in a post-normal world.

A post-normal world needs a guiding ethic. In Chapter 8, lawyer and environmental activist, Dianne Dibley, accepts the current label for our epoch as the Anthropocene, a time of disruption of the planet's life-support systems. She proposes the intersection of science, economics, law, ethics and common sense as the source of an environmental ethic for the future. The convergence of these different forms of knowledge creates a shared space for ethical decision making; a reconciliation space that the ancient Greeks called *sophia*.

In Chapter 9, for Charles Massy, the author of *Call of the Reed Warbler: A New Agriculture – A New Earth*, the dilemmas of the Anthropocene have arisen from a global shift from an organic to a mechanical mind. Inherited from each country's first peoples, the organic mind is in tune with the rhythm of the natural environment. The mechanical mind limits thinking to the physical world. As a result, agriculture has been confined to machines that strip the environment and imperil the future. Charles recognizes the rise of an open and creative emergent mind, one for whom a fresh environmental ethic is already leading to the spread of regenerative agriculture around the world.

Regeneration is a word that is coming into its own. It follows a proud tradition of a human capacity for transformational change. The 14th-century Renaissance was a rebirth, not a repetition, of the classical design, art and science of Ancient Greece. The 15th-century Reformation erupted as shift from institutionalized religion to the role of the individual. Regeneration appears to be following a similar path, respecting the past and embracing the new. Charles Massy suggest that this goes beyond regenerative agriculture to a regenerative society. In Chapter 16, David Marsh describes what it means to be a regenerative farmer.

During research in Northern Australia, Kate Andrews, a research associate of Charles Darwin University, discovered the cycle of events that happens when introducing new ideas. In Chapter 10, her case studies of many years of failed cropping revealed that new initiatives reverted to the old ways, to be rediscovered over and over again. Her research uncovered the driving factors that kept the circle turning. She called this impasse a circular conundrum. Kate concluded that the driving factors in this cycle are the outcome of the dominant culture and values of the country and they need to be changed if the cycle is to be broken.

In another study of attempts to change mindsets, the change was the worldview of veterinarian and poet, David Waltner-Toews (Chapter 11). He describes how, working in poverty-stricken Nepal, he changed from his identity as an empirical scientist to that of a post-normal science practitioner. This led him to treat everyone in the community as responsible individuals, rather than units in an epidemiological study. Within three years those same individuals were working together to create a complex, multilayered, cooperative society, of which, they said, their grandchildren would be proud.

The power of an idea eventually rests on the power of the action that follows. Philosopher David Hume moaned that his 18th-century *Treatise of Human Nature* 'fell deadborn from the press'. This became the basis for a whole school of philosophy. In the preceding chapters, we were exploring powerful new ideas. In the chapters that follow, new ideas, including regenerative farming, have been applied in practice.

The division between ideas and practice is always arbitrary. The ideas of the Fenner Writing Group members have a practical foundation. The ideas put forward in the previous chapters were post-normal reconciliation among interests, an environmental ethic of including all knowledges in *sophia*, an emergent mind that could remain open minded in processing new ideas, the circular conundrum of hope and then failure and the emergence of regenerative thinking. In the chapters that follow, independent thinkers have put those ideas into practice.

In Chapter 12, Richard Thackway and Bruce Gardner, a physical scientist and a social scientist, describe how they monitored two regenerative farms for their production, profitability and lifestyle. Both farmers had shifted from a mechanical to an emergent mindset. The evaluation was based on objective data, as well as data on social and ethical aspects. In interviews, the farmers had distinguished between 'maximum' profit and 'enough'. 'Maximum' was based on the use of herbicides and industrial machinery. 'Enough' meant sufficient to meet environmental, family, social and personal needs.

A significant influence on ideas about language, the Sapir-Worf hypothesis, claimed that: 'The limits of my language form the limits of my world.' This epigram presents a challenge to find a new language for new mindsets. Craig Ashhurst, an organizational change consultant, was working on the wicked problem of institutional change. In Chapter 13, he describes how he found that each of the interest groups involved in an organizational change had developed a shared language, ideas and aspirations: a collective coherence.

Bringing together multiple interests has been called transdisciplinarity. However, this does not convey a coherence among coherences. Craig coined the word transcoherence to take account of a system of coherences that arises in transformational social change.

In any discussion of social change, a major question is who holds the power. In the 1980s, care organizations were being asked to change from hospital-type services to a caring-community model. In theory, their clients therefore changed from being a stereotype of disadvantage to being individuals in their own right. In practice, the power lay with the service providers, the clients felt powerless and the reforms lay unresolved.

Sociologist and change agent, Julia Wolfson, working with institutions around the world, identified another source of power, the deep powers within every individual. In Chapter 14, she describes how she is basing her practice on helping clients, clinicians and administrators of care services to release their deep powers.

Research has long been considered a matter for the individual researcher. Longstanding dominance of empirical science, and a culture bent on polarized debate have kept the individual being considered the originator of new knowledge. Yet scientific practice rejected the subjectivity of individual collection of evidence. Any reference to 'I' was banished from scientific reports.

Rebecca Freeth, Elizabeth A. Clarke, and Dena Fam address this paradox from their own experience in collaborative research. In Chapter 15, they write of their joint study of research practice, which revealed that researchers considered their work as being from an 'I' that they could not even acknowledge in their reports. Things have now changed. A researcher is now required to work in teams and to acknowledge her institution. The authors noted that developing these relationships as 'we' is seldom regarded as part of the inquiry process; much less given their own status as a distinct mode of inquiry. The proper definition of this inquiry method recognizes a research team as 'we'.

A theme that resonates throughout the writing group chapters is regenerative practice. More usually linked to working with the land, the term is also applied to social change and to the individual thinking for themselves. David

Marsh, who describes himself as a regenerative farmer, writes of regenerative practice as 'allowing', allowing lands and communities to self-organize and self-repair, to move on from the mechanical mind to a steady state that encourages diversity.

For an individual, regeneration means to allow oneself to stand back and let the system of which one is a part to self-organize in its own best interests, and hopefully in yours. John Seed described this form of thinking as: 'I am not saving the rainforest, I am part of the rainforest saving itself.'

David Marsh put this into practice by respecting the land's capacity to look after itself, ensuring the time and independence for it to do so. Under this regime, David reports better health of his land, stock, crops and family; equal or better profits; resilience in hard times and free recreational time.

This story is about the generation of ideas, followed by the development of their practice, 'walking the talk' as it is described. The story created from the chapters written by the members of the Fenner Writing Group offers the potential for an epoch in which human beings are reconciled with each other and with their environment: a post-normal society. As well as the over-arching story of regeneration, there were other avenues through which the chapters came together.

In Chapter 2, in a series of workshops on whole-of-community change members were asked their priority issues for the future of their society. The issues were climate change, employment, security, confident youth and healthy aging, all issues for a community's future. Workshop members were then asked their priorities for themselves, which produced a very different list: identity, reconciliation, social learning and social change, language, personal power and an independent mind, all issues associated with independent thinking.

While the thinking dimensions do not stand alone, it is often possible to identify the entry point by which each individual accesses the full power of their mind. In the chapters that follow, it is possible to identify an entry point, which may or may not have been the intention of the author. It is important to note that whatever the entry point, the full set of dimensions would always be in play.

Moving through the individual dimensions of thinking in turn, Thackway and Gardner drew on the physical dimension, before expanding into the social (the farmer), the ethics of regeneration, the aesthetics of a landscape and sympathy between the farmer and his land. Andrews emphasized the physical status of cropping in Northern Australia. Authors concerned with social learning entered through the social dimension, Arabena through reconciliation, Ashhurst through language and Waltner-Toews through fieldwork as a veterinarian.

Others used the ethical dimension as the basis for their chapter, Dibley searching for an environment ethic for this epoch and Wolfson for the independence for individuals in care. The aesthetic dimension is the one often least made explicit. In these chapters, Massy and Marsh capture the reader through their sense of the beauty of the landscape. The sympathetic dimension is entered through Freeth, Clarke and Fam's interest in teams and Marsh's identity with the land.

With such diversity in so short a space, it is difficult to sum up the contribution of the writing group authors to independent thinking in this uncertain world. The story of the tapestry of their contributions is one way to sum up. Their choice of topics, their knowledge base and their favoured dimension of thought are other ways of knowing that we have pursued here.

There is yet another way to review the rest of Part II. At the beginning of this chapter, we drew attention to popular, and mistaken, beliefs about independent thinking. This was a view from the 20th century. For the 21st century and beyond, in an uncertain world the holistic contribution of independent thinking becomes essential. So one further question that one can ask of the chapters in Part II: to what extent do they provide prototypes and frameworks for the multidimensional thinking of tomorrow?

The chapters in Part II are diverse, individual and creative. Their authors have developed a language, frameworks and methods of inquiry that enhance multidimensional thinking. Their chapters, taken separately, address many of the issues that bedevil an increasingly uncertain world. For those seeking to follow in their footsteps, they represent a valuable resource.

Taking the range of prejudices against independent thinking as **Reality 1**, let us consider the case where the opposite realities are true. Together, the second set of realities, confirmed throughout this book, provide criteria for recognizing independent thinkers.

Applying the 'which reality can you trust' test to the each of the chapters in Part II:

Reality 1. Independent thinking is a rare event. **Reality 2.** Independent thinkers consciously draw on the multiple dimensions of their minds.

Reality 1. Independent thinkers create unresolvable tensions within a group. **Reality 2.** Resolution of tensions is part of the skill set of every independent thinker.

Reality 1. Individual thinkers reject the thinking of others. **Reality 2.** Independent thinkers will consider all possible alternatives to the existing system.

Reality1. Independent individuals can help resolve the prevailing issues. **Reality 2.** Independent thinkers will question the prevailing description of an issue, since it is likely to be part of the problem.

In a world that questions even the existence of independent thinking, the chapters in Part II provide an antidote. The lessons to be found in each chapter that follows and the criteria provided by previous chapters offer a firm platform from which independent thinkers can recognize themselves and even enhance their practice.

Before moving on from these ways to recognize independent thinkers, to the chapters themselves, perhaps Aldo Leopold sums it up best:

> Examine each question in terms of what is ethically and aesthetically right, as well as what is economically expedient. A thing is right when it tends to preserve the integrity, stability and beauty of the biotic community. It is wrong when it tends otherwise. (Leopold, 1949, pp. 224–225)

Bibliography

Hume, D. (1902) *Enquiries Concerning the Human Understanding and Concerning the Principles of Morals by David Hume, 1777*, Selby-Bigge, L. A. (ed.), 2nd edn. Oxford: Clarendon Press.

Leopold, A. (1949/1981) *A Sand County Almanac and Sketches Here and There*. London: Oxford University Press.

Massy, C. (2017) *Call of the Reed Warbler: A New Agriculture – A New Earth*. St Lucia, Australia: University of Queensland Press.

Ravetz, J. (1999) 'What is Post-Normal Science?', *Futures*, 31(7), pp. 647–653.

Ravetz, J. (2005) *A No-Nonsense Guide to Science*. Northampton: New Internationalist.

Whorf, B. L. (1959) *Language Thought and Reality*. New York: John Wiley & Sons.

Part II

Practice

7

Post-normal reconciliation

Reframing the agenda

Kerry A. Arabena

The ideas I want to express here are provocative and I take my role as a provocateur very seriously. Before I get to the body of the chapter, I wish to tell you how I came to appreciate and use the sciences, particularly those in the ecological tradition. I came to adopt the form of science that includes aspects of uncertainty in the framing of complex issues, referred to as post-normal science. Until recently, I was accepting of our current Australian-based reconciliation activities – what I refer to here as 'normal reconciliation' – but I will describe some of the regional uncertainties that made me decide that we need to look beyond normal reconciliation. I refer to this new body of work, which is attentive to the lessons from sciences of many cultures as 'post-normal' reconciliation; an extension or a 'beyond' what we are currently doing, not because it is interesting, but because we *must*. I will then touch on what Earth system, environmental and quantum sciences have to offer this agenda and show where there are synergies with the sciences of indigenous peoples from three continents across the world. This is a body of work that will probably interest me for the rest of my life. Even as I set out a new way of viewing reconciliation, I concede it is beyond me to also offer the strategies needed to get there or how this new agenda is to be enacted. I firmly believe that provocation is only one side of the conversation; people need to be able to respond.

Responses happen after an act of listening. So, in order to better respond to the ideas presented here, I ask you please to keep an open, relaxed mind. It is your imagination that I am speaking to. To evoke your imagination, it is imperative that we suspend ourselves from the normal encumbrances that we use to anchor ourselves to our everyday comfort, to relationships, to identity, to society. It is your inner self that I seek to engage in this conversation.

For it was my inner self that was stirred through engaging with science. I maintain a respect for the contributions of traditional Western science. I became keenly attuned to the work of Earth system, environmental and ecological scientists who have tested new waters and are often only tentatively accepted by the established scientific community. Some of these scientists are themselves concerned with challenging the assumptions that have underpinned traditional scientific views.

Others are finding increasing unison with the voices of indigenous peoples and ecological philosophers who are highlighting that achievements in science, technology, industry, commerce and finance have brought humans into a new age at the expense of much of the diversity of life and the life-enhancing processes of the Earth. From within these sciences, we find ways to be concerned with and celebrate the importance the integrity and the mystery of diversity of life in all its forms, in the entire world. The diversity in the Arctic and tropical regions, the oceans and continents, mountains and valleys, forests and deserts, rivers and their floodplains all give Earth its endless wonder and its functional integrity. Drawing on these sciences, we might say that the planet is a singular reality composed of diversity beyond all our understanding and description. How I came to be engaged with science is this. After two decades of working in Aboriginal and Torres Strait Islander health, I gained a position which provided stewardship for the largest collection of Aboriginal and Torres Strait Islander artefacts, language recordings and documents in the southern hemisphere at the Australian Institute of Aboriginal and Torres Strait Islander Studies (AIATSIS). Working in this position, I became saddened by the disappearance of cultural and linguistic knowledge and biological diversity in Australia. This loss has been precipitated by the comparatively recent colonization of Australia's First Peoples and land. Prior to colonization, there were over 300 languages in existence. It has been stated by leading ecologists that linguistic diversity corresponds to a living diversity of cultures and ecosystems. This said, we can deduce that there were over 300 ecosystems in Australia and the knowledge and languages that maintained those ecosystems in a state of equilibrium for over 50,000 years, held by the First Peoples, have been erased through modern Australian knowledge and political systems. This was a powerful awakening for me, one that I could not ignore, even if I tried.

I started looking for questions and answers to find ways by which to communicate our indigenous peoples' role in caring for the diversity of the planet and how this worldview had been reduced to a collection in AIATSIS. I thought about the interaction of indigenous peoples across the world at war not with humans per se, but with human thoughts and actions that reduce the planet to a place subject to human control. I started to seek the questions and answers that could make this conversation possible. Interestingly, I did not find them in the humanities – humanist scholars tell us this school of thought provides for the expansion of humans' quality of life. Yet this centring of value is so extensively human it distorts the place and role of humans in ecosystems.

There are very few places in the academy where I could openly discuss the relationship between the health and wellbeing of humans and other species *and* promote the fact that human health is dependent on healthy functioning ecosystems and all the life those systems contain. I found the questions and answers I required in human ecology, a science that ties humans to nature in ways that promote a reciprocal relationship – we are touched by what we touch. We are shaped by what we shape. We are enhanced by what we enhance. This is the basis of indigenous peoples' sciences, and evident in Earth system, environmental and ecological science.

All of these sciences combined have given me an appreciation of the physical processes at work in the universe as never before and they have also given me a profound critique of the devastation wreaked by human beings. This is particularly true of those of us who live in industrialized nations around the world. I have reviewed literature that describes the entire universe as an interdependent system. Earth system scientists and physicists working on the Intergovernmental Panel on Climate Change and the Intergovernmental Geosphere-Biosphere Program along with people's movements are promoting the transition from the scientific technological phase into the ecological phase of the human–Earth relationship. And, being true to my custodian of country heritage, I am compelled to go with them. They have provided enough evidence – we have to be terrified by what we have done, but not without hope.

At the commencement of the 21st century, I am mindful of the fact that we are looking to secure our First Peoples' economic, environmental, social and cultural futures in an unprecedented period of change in the world. We are losing biodiversity and have damaged our waterways. A great number of species are facing extinction.

The holders of the oldest ecological knowledge in the world are being forced to participate in education, political and modern systems that have little regard for that knowledge and almost always no regard for their country. When we lift our gaze beyond this shore, we see that we have the largest number of adolescents in the history of our species, many of whom are living in poverty. India and China are becoming the most powerful technologically advanced countries in the world. Globalized, multinational companies have larger budgets than some countries. Our country is at war, our region in conflict. That we are in the midst of a crisis is now well understood. Not only does the reconciliation agenda have to do with indigenous people and non-indigenous people, but with people who support us and the country that supports us. Reconciliation should also concern Australia's place in the Asia Pacific region, so profoundly are our futures entwined.

I think we all know that, in order to truly reconcile, we need to take stock of where we are, who we are and where we want to be in 100 years' time. We need to start with the story of Australia how it and its people came to be, we need to reconcile our hopes for the future within this country and with others in the Asia Pacific region. We need a long-term and inclusive view, we then need to purposefully craft a new story around that view and use that story to build new ways of living, new ways of being and new ways of having relationships. It is an extension of the reconciliation agenda we have in Australia, a view 'beyond' where we are now. The realization of this story, and our collective commitment to achieving the aspirations inherent in this story, is the vision for my post-normal reconciliation agenda. A post-normal reconciliation agenda is only possible because of our efforts to achieve reconciliation to date. It is our thinking and imagination that makes such reconciliation possible. The realization of reconciliation was an idea, a concept a few years ago. And we have nurtured it along to what it is today. Ideas are at the heart of this realization.

Let me speak of what normal reconciliation is. Reconciliation as a term, as a legacy, as an experience, is largely dependent on acts of forgiveness. Who can forget the truth and reconciliation councils in South Africa? Who can forget the Australian Bringing Them Home Report that captured the stories of separation and pain of Aboriginal children, of their resilience and of the nation's apology? Reconciliation as a term is transformative, oft used as a tool of negotiation and peace making. It is different to other negotiation or peace-making strategies. The essence of reconciliation is the voluntary initiative of the conflict parties to acknowledge their responsibility and guilt. The interactions that transpire between the parties are not only meant to communicate one's grievances against the actions of the adversary, but also to engage in self-reflection about one's role and behaviour in the dynamic of the conflict. The aim of such interaction is that, in the final analysis, each of the parties acknowledges and accepts his or her responsibility and, out of such recognition, seeks ways to redress the injury inflicted, to refrain from further damage and to construct new positive, relationships.

In Australia, normal reconciliation seeks to improve the lot of Aboriginal and Torres Strait Islander people and is stated as the country's unfinished business. Reconciliation as a practice has been rightfully constructed as a social norm – to recognize and give value to First Peoples, to understand the meaning of our cultures and to foster stronger relationships between mainstream and indigenous Australia. These are important goals and have become important to a great many people. Reconciliation Action Plans, the Sea of Hands initiatives, the Sorry

Books, the Bridge Walk, Journey of Healing groups, and Reconciliation Week are heartfelt activities that have garnered support for reducing the disadvantage experienced by Aboriginal and Torres Strait Islander people and work to build mutually respectful relationships between indigenous and other Australians. These relationships allow us to work together and generate success that is in everyone's best interests.

Achieving reconciliation in Australia has involved raising the awareness of, and knowledge about, indigenous history and culture; changing attitudes that are often based on myths and misunderstandings and encouraging action where everyone plays a part in building a better relationship between us as fellow Australians. Reconciliation Australia's ambition is to eliminate the glaring gap in life expectancy between indigenous and other Australian children. The organization states that the life of an indigenous child is likely to end significantly earlier than that of a non-indigenous child's. It has stated that this is not acceptable in an affluent country like ours. And I agree.

Now, while these 'normal' reconciliation activities are critical, the spaces in which reconciliation occurs is more often than not structured by humans and about human concerns. 'Normal' reconciliation tends to be enacted within policy, in organizations, in gatherings of humans, outside parliaments, around meeting tables, in schoolrooms, on streets and in legislation and in protests and poems.

Rarely does reconciliation happen in the natural environments that Aboriginal and Torres Strait Islander people have asserted time and again are the mainstays of our culture. Aboriginal and Torres Strait Islander people have asserted that we are not separate to country, that we are imbued with it, we have responsibilities for it even when we are not born on country; it is our homeland, our heartland, filled with songlines and alive for us in ways that are often beyond the comprehension of people who are not profoundly connected to or from country. If normal reconciliation activities were to happen in the natural environment or in a First Peoples' context, then the activities and ways of participating in reconciliation would, I imagine, be different. Could we have a Sea of Hands in a forest? A Reconciliation Plan in ecosystems, on country? Would we have a Bridge Walk of a million people across rivers and wetlands? Would we ask for democratic rights in the middle of a ceremony acknowledging a young person's coming of age?

No? Then what would we do together, in the process of reconciliation in a First Peoples' context? While 'normal' reconciliation aims to displace the privilege of mainstream Australia through democratic means, it is difficult for Aboriginal and Torres Strait Islander people to engage fully in the language for and about reconciliation, because it is difficult not to react or be conciliatory to the 'colonial project' in this country. I was asked: 'Kerry, how can I reconcile with them when they have all the money and all the best land and I have nothing?' Another told me that she felt hated and asked how can we reconcile with people who hate us. In the oft-used language of reconciliation, Aboriginal and Torres Strait Islander people still need to 'close the gap', to 'overcome our disadvantage', to 'be successful' in ways that structure our entry into modern Australian society first as 'clients', then as 'independent and valued' citizens of Australia, then as 'consumers'. We are to want what everyone else wants – to take up our rightful place in a mainstream middle class and enjoy all the benefits and resources that Australia and credit card companies have to offer.

What is problematic about positioning First Peoples in this lineal trajectory toward a reconciled Australia is how we are viewed and how we have adopted that view of ourselves in relation to everything else. Understanding this view led to me ask questions in an effort to reframe where we are at and where we might go to with reconciliation in the future. Those questions were: What would happen if we removed indigenous and non-indigenous categories; what would we be? Why are we so heavily invested in these descriptors? What would happen if we took the view that the sickness that permeates our indigenous society is a symptom of the way life is lived in modern Australian society? Is there a relationship between these two things, modern wealth and indigenous disadvantage? What would happen if the disconnect between modern Australian society and the ecosystems in which they have arisen is so profound, that the country is so raped, pillaged and plundered that, because of our sensitivity to country, we have no choice but to be sick? What is it that the rest of Australia has to give up in order to gain a relationship with Aboriginal and Torres Strait Islander people? What is it *we* have to give up? And what is there to gain? What do we have to be open to in order to change?

These reframing questions and their answers move the experience of reconciliation out of the colonial project into new spaces and new opportunities. This is the space that can lend itself to post-reconciliatory discussions, holding onto parts of our reconciliation story that are instructive, but opening to new conversations, creating new ideas and new realities.

Now, remembering that reconciliation should engage us in self-reflection about one's own role and behaviour, that we have reframed reconciliation to look beyond our current normal reconciliatory practices to take a long-term view and that we have shifted reconciliation outside the colonial practice by moving into a First People's context, that context being land, we do have a new space for post-normal reconciliation, not only to reconcile between indigenous peoples and fellow Australians, but to reconcile us all with the natural systems in which we live and to reconcile with other species with which we spend our time on Earth.

The vision for a post-reconciliation agenda should be that the full fluorescence of the world in all its magnificence is here for future generations. In normal reconciliation, we will carry out the improvement in the circumstance of our lives and in post-normal reconciliation, we will be part of the global infrastructure that facilitates a transition from a period of human-instigated devastation to a period when all humans live on the planet in a mutually beneficial manner. This is the natural trajectory for reconciliation, a fluid movement between reconciling people, to reconciling people and the planet.

Such a process of reconciling people with the planet has no human historical parallel. Earlier humans have not dealt with anything comparable to the toxins in the air, the water, the soil or with the immense volume of chemicals dispersed throughout the planet. Neither were they dealing with the extinction of species or the altering of climate on a scale such as is our present concern. In order to ensure the planet is available for future generations, we have to reconcile our older knowledge traditions with our modern ones and reconcile our views about then act on how to live within the very structure and functioning of the planet itself.

This transition is difficult to be sure; however, there is precedent in thinking how to achieve this. In 1962, Rachel Carson published the book *Silent Spring* that played a large role in articulating ecology as a subversive subject – as a perspective that cut against the grain of materialism, scientism and the technologically engineered control of nature. The book argued that uncontrolled pesticide use was killing not only animals and birds, but also humans. Its title was meant to evoke a spring season in which no bird songs could be heard, because they had all vanished as a result of pesticide abuse. She made it clear she was not advocating the banning or complete withdrawal of helpful pesticides, but was instead encouraging responsible and carefully managed use, with an awareness of the chemicals' impact on an entire ecosystem.

The ideas encapsulated in post-normal reconciliation make possible the new ground in the new stories needed for our time – the myths of our cultures, our paradigms, our beliefs – that form the core of what we tell ourselves is reality, both now and in the future. Stories, in this context, are anything we add to our original experience that alters what we think is going on or changes how we think about things. From 'We need electricity to be comfortable and maintain our way of life' to 'Producing electricity is pumping billions of tons of carbon dioxide into the atmosphere, leading to global warming and extremely destructive weather patterns.' From 'Being able to drive where and when we want at a low cost is freedom!' to 'Our driving habits are feeding the destruction of the planet.' From 'All of nature is here to serve mankind' to 'Humans are no more or less important to the planet than any other life form.'

In fact, every time a culture has been transformed – from what it was before and for better or worse – it has been because of an idea, an insight, a new understanding of how things are and what is possible. Ideas precede every revolution, every war, every transformation and every invention. And while it was in the political and social sciences I found the language with which to build the story, I have used the environmental, earth science and quantum physics stories of ourselves, our planet and the universe as the inception point from which to build a post-normal reconciliation agenda.

Quantum physics (the new physics) describes the planet as forming part of a highly integrated, coherent system that implies that at a deep and fundamental level the seemingly separate parts of the universe are connected in an intimate and immediate way. Lazslo developed a theory in which different things at different locations in space and different points of time are linked – particles, atoms, molecules, organisms, ecologies, solar systems, entire galaxies, as well as the mind and consciousness associated with some of these things – regardless of how far they are from each other and how much time has passed since these connections were created between them. Capra felt that concepts in physics have the potential to render a profound change in our worldview, from the mechanistic conceptions of Descartes and Newton to a holistic and ecological view, similar to that of indigenous peoples, mystics and sages of the ages.

Our new story, the post-normal reconciliation dream, is premised on the fact that we exist in an interconnected life system. Indigenous philosophies show that it is possible to learn to be in the world in reciprocal relationships with all things, through cooperation and constraint, interdependent thinking, morality and action. Scientific evidence indicates that our planet is a one-time endowment, profoundly interrelated with the constant energy source provided by the sun and now, more than at any other time in history, humans need to understand and respond effectively to our own human role given this new information.

Because of this scientific evidence, from indigenous peoples, Earth system sciences, quantum physicists and environmentalists, I ask you to imagine this: There is no difference between indigenous peoples and non-indigenous peoples. We all belong to the human species. There is no difference between the knowledge of indigenous peoples and that of non-indigenous peoples. It is all *human knowledge*. There is no physiological difference in our brains in how they function. We *generate* human knowledge. There is evidence to demonstrate conclusively that there is no separation between us as people and there is no separation within the natural systems in which we live. We are already, at a biophysical level, reconciled.

What is different is the energy context that gives rise to that knowledge. Modern Australia's knowledge is founded on fossil fuel. Indigenous peoples' knowledge is founded on sunlight. The stability of sunlight as an energy source has seen the capacity of indigenous people to hand down knowledge of how to live in country, generation after generation. Petroleum as an energy source makes us oblivious to our natural environment. We are at a point at which we can transition from this petroleum-based energy context and it is a moment of significance far beyond what any of us can imagine.

Understanding ourselves as deeply and profoundly connected means that there is no one race that can lay claim to indigenous wisdom; it lives deep in the heart of every living creature, anyone who remains deeply aware of the rhythms of the natural world can remember it. I can say this because the languages and beliefs of our collective ancestors were not embedded in colonial thinking and were not used to construct privilege. And because of this, I urge you to imagine this – we are all indigenous, to this universe. We are already reconciled, not only between each other, but with every other being with which we live our lives. And as such, we all have custodial responsibilities, we have to reify the ecosystems in which we live, make subordinate our human individual requirements and practise personal development by responding to the needs of our whole community – including those of other species. This is task of the post-reconciliation agenda.

What we have to reconcile then is a set of complexities and tensions that are becoming the primary tension in human affairs. We have to reconcile developers and ecologists on the state of our natural world. The economic tensions between empires and colonies are shifting to being between the village peoples of the world with their organic way of life and transnational corporations with their industrialized agriculture. What we have to reconcile belongs no longer in divisions based on political party or social class or ethnic group. What needs to be healed is a division based on humanity being one of the components within the larger community of the planet Earth. We cannot mediate the present situation as though there were some minimal balance already existing that could be slightly modified – the violence already done is on a scale beyond acceptability. And again, indigenous peoples in Australia are most vulnerable to the effects of these tensions, as is the land for which we care.

To make this post-normal reconciliatory programme a reality, we need new tools for new problems. One such tool is also found in the sciences – post-normal science is a method focusing on aspects of problem solving that tend to be neglected in traditional scientific and indeed in policy development: uncertainty, value loading and plural legitimate perspectives. Post-normal science includes these uncertainties in the framing of complex issues and provides an opportunity for extending more and different people's participation in decision making. The shift to post-normal science or other methods for dealing with uncertainty and complexity is a necessity.

To adopt such a method is a critical change in this country because it allows the emergence of new strategies in which the role of indigenous science and new sciences are appreciated in their full context. We need these methods to infiltrate not only our economic system, but government, jurisprudence, the medical profession, religion and education. Industry in the interests of promoting human development is producing a recession of the most basic resources of Earth. We must adapt and adopt new ways of living. We never thought of ourselves as capable of doing harm to the very structure of the planet or extinguishing the living forms that give the world its grandeur. So much now depends on our decisions about our way into the future. So, what's it going to be, my people?

I have offered my ideas to both provoke and stimulate. I have discussed the need to expand the current reconciliation agenda, established between First Peoples and other Australians, into its broadest articulation, into a post-normal reconciliation agenda. I do this not to denigrate what has been achieved to date in current reconciliation practice or to say that our work in this realm is finished. It is not. Beyond these actions, there is an expanded reconciliation agenda, because we all have a role and a responsibility to protect the health of ecosystems and all things that reside in them. I have written about the synergies between indigenous peoples' science and the science of Earth system scientists, environmentalists and ecologists and the need to build a reconciliation agenda not between people, but between people and living systems.

It is these interdependent systems that give us life, that give us reverence, give us a story in which we can all celebrate nature and the deep mystery of things in a meaningful way. We have a chance to change the soul of the modern world, not just technologically, not just to get higher wages or to even get physically improved conditions for indigenous peoples, but to change our inner world, to have a vision of a world transformed through stories that we created to positively impact on each other and the planet.

All that I have offered here has been offered constructively and in a spirit of reconciliation and is a celebration of the greatest articulation of our lived experiences. I have hope in my heart about the future my people and this hope is for us all. Because when the world quiets to the sound of our own breathing, we all want the same things: comfort, love and a peaceful heart.

Acknowledgements

I wish to pay my respects to my Elders, and to thank Matilda House for her generous welcome to country. Some very special non-indigenous people have helped me both personally and professionally: Professor Emerita Valerie Brown and participants in the Human Ecology forum of the Fenner School of the Australian National University who introduced me to science in a way that made my worldview permissible in the Academy.

Bibliography

Alfred, T. Peace (1999) *Power and Righteousness: An Indigenous Manifesto*. Oxford: Oxford University Press.

Alfred, T. Peace (2005) *Wasase: Indigenous Pathways of Action and Freedom*. Toronto: Broadview Press.

Alfred T. Peace and Corntassel, J. (2005) *Being Indigenous: Resurgences against Contemporary Colonialism', in Government and Opposition*. Melbourne: Blackwell Publishing.

Arabena, K. (2005) 'Not Fit for Modern Australian Society: Aboriginal and Torres Strait Islander people and the new arrangements for the administration of indigenous affairs', in *Australian Institute of Aboriginal and Torres Strait Islander Studies Discussion paper*, No. 16.

Arabena, K. (2006) 'The Universal Citizen: An indigenous citizenship framework for the twenty-first century', *Australian Aboriginal Studies*, 2, pp. 36–46.

Arabena, K. (2008) 'Editorial: "I, Torres Strait Islander"', *Zenadth Kes Torres (Strait Islander Journal). Thursday Islander*, 1, pp. 1–4.

Arabena, K. (2008) 'Indigenous Epistemology and Wellbeing: Universe referent citizenship', in *Australian Institute of Aboriginal and Torres Strait Islander Studies Discussion Paper*, Number 23.

Arabena K. (2008) 'Policy Imagination: The possibilities of synthesis for eco-health practitioners', *Zenadth Kes Journal (Torres Strait Islander Journal). Thursday Islander*, 1, pp. 25–33.

Arabena, K. (2009) *Indigenous to the Universe: A Discourse of Indigeneity, Citizenship and Ecological Relationships*. Unpublished thesis, Fenner School, Australian National University.

Arabena, K. (2009) 'We Are All Indigenous to the Universe', in K. Arabena and C. Fforde (eds.), *Giving an Account of Ourselves: Aboriginal and Torres Strait Islander Contemporary Philosophy and Worldview*. Canberra, AIATSIS Research.

Arabena, K. (2010) 'All Knowledge Is Indigenous', in V. A. Brown, J. A. Harris and J. Y. Russell (eds.), *Tackling Wicked Problems: Through the Transdisciplinary Imagination*. New York: Earthscan.

Arabena, K. (2015) *Becoming Indigenous to the Universe: Reflections on Living Systems, Indigeneity and Citizenship*. North Melbourne: Australian Scholarly.

Basso, K. H. (1996) *Wisdom Sits in Places: Landscape and language among the Western Apache*. New Mexico: University of Mexico Press.

Bateson, G. (1973) *Steps to an Ecology of Mind: Collected Essays in Anthropology, Psychiatry, Evolution and Epistemology*. St Albans: Paladin.

Bateson, G. (2002) *Mind and Nature: A Necessary Unity*. New York: Hampton Press.

Berry T. (1999) *The Great Work: Our Way into the Future*. New York: Bell Tower Press.

Berry, T. (2006) *Evening Thoughts: Reflecting on Earth as a Sacred Community*. M. Tucker, (ed.). San Francisco: Sierra Club Books.

Berry T. and Swimme, B. (1992) *The Universe Story*. San Francisco: HarperCollins.

Bohm, D. (1980) *Wholeness and the Implicate Order*. London, Routledge.

Bohm, D. (1993) The *Undivided Universe: An Ontological Interpretation of Quantum Theory*. London and New York: Routledge.

Bookchin, M. (1980) *Toward an Ecological Society*. Montreal: Black Rose Books.

Bookchin, M. (2005) *The Ecology of Freedom: The Emergence and Dissolution of Hierarchy*. Edinburgh: AK Press.

Brown, V. A. (2008) *Leonardo's Vision: A Guide to Collective Thinking and Action*. Melbourne: Sense Publishers.

Callicott, J. B. (1989) *In Defence of the Land Ethic: Essays in Environmental Philosophy*. New York: State of New York Press.

Carson, R. (1962) *Silent Spring*. Boston: Houghton-Mifflin.

Deloria, V. (2003) *God is Red: A Native View of Religion*. 30th Anniversary Edition. Golden: Fulcrum Publishing.

Gumurdul, D., Mikginjmikgin, P. and Nayinggul, J. (2006) 'Open Letter to the Prime Minister and People of Australia', *The Age*, Melbourne, 8 August.

Lazlo, E. (2007) *The Systems View of the World: A Holistic Vision for Our Times*. New York: Hampton Press.

Lovelock, J. (1990) 'Hands Up for the Gaia hypothesis', *Nature*, 344, pp. 100–102.

Margulis, L. (1988) *Symbiotic Planet: A New Look at Evolution*. London: Weidenfeld & Nicholson.

Shiva, Vandana (1989) *Staying Alive: Women, Ecology and Development*. London: Zed Books.

Shiva, Vandana (1997) *Biopiracy: The Plunder of Knowledge and Nature*. Cambridge, MA: South End Press.

Steffen, W. P., Crutzen, P. and McNeill, J. R. (2007) 'The Anthropocene: Are humans now overwhelming the great forces of nature?', *Ambio*, 36(8), pp. 614–621.

Vernadsky, V. I. (1945) 'The Biosphere and the Noosphere', *Scientific American*, 33(1), pp. 1–12.

Zukav, G. (1979) *The Dancing Wu Li Masters: An overview of the new physics*. Richmond: Rider/Hutchinson of Australia.

8

Sophia in the Anthropocene

Towards an environmental ethic

Dianne E. Dibley

Introduction

In the not too distant past we entered a time in Earth's natural history in which the superabundance of human beings and their power to shape the biosphere exploded into planet-scale changes (Pugh and Renner, 2014). The idea of the Anthropocene as a new epoch in geological history was first proposed by Nobel Prize-winning atmospheric chemist Paul Crutzen in 2000. It has since been widely adopted by both the scientific and general communities.

However, in July 2018 the International Union of Geological Sciences (IUGS) made an unexpected decision not to recognize the Anthropocene on the geologic time scale. In making this decision, the IUGS ignored the recommendations of its own Anthropocene Working Group (AWG). In a paper published in 2016 in *Science* magazine, the AWG had argued that the *Anthropocene is functionally and stratigraphically distinct from the Holocene and* the changes are as big as those that happened at the end of the last ice age. They claimed that these changes have driven the Earth out of the Holocene epoch in which agriculture and civilizations developed. In recognizing the new epoch, the AWG claimed that the Anthropocene does have a robust geological basis and is in widespread use. Indeed, it is becoming a central, integrating concept for understanding global climate change (Waters et al., 2016).

The decision to regard anthropogenic changes to Earth systems as being within Holocene variability has been said to reflect internal politics in the IUGS. Scientists Maslin and Lewis (2018) claim that what happened behind the scenes was a race between two 'committees', with some scientists wedded to the Holocene and others backing the acknowledgement of the Anthropocene. They further claim that it is owing to the competitive committee structure that presides over the geologic time scale that has caused the rancour in scientific circles.

Whether or not the Anthropocene is formally recognized on the geologic time scale by the IUGS, it has significant scientific and general currency – it is part of the *lingua franca*. Maslin and Lewis (2018) argue out that it seems naïve to define an age we live in without mentioning the geological-scale impact of humans on the Earth system. They continue that the Anthropocene rewrites the human story, highlighting the need for planetary stewardship.

Even if the Anthropocene fails to be formally recognized on the geologic time scale, I would argue that it has legitimacy as an epoch in Earth's natural history. This is the epoch in which my enquiry takes place.

The role that humanity has played in the environmental crises of the Anthropocene and our continuing failure to take effective action raises questions about how we think individually and collectively. There is a growing body of literature from across disciplines arguing that we need to reinstate common sense to our decision making (Sachs, 2011). I argue that the pre-eminence of rational thinking has left humanity stranded in an imploding world of its own making (Dibley, 2018).

Recognition that different types of knowing go together to make up an individual's or indeed a society's knowledge set is documented in the literature. For example, Aristotle identified different types of knowledge – *episteme* (conceptual), *techne* (technical) and *phronesis* (practical). To this, he added an overriding knowledge – wisdom, which he called *sophia*. Polanyi's (1958) picture of knowledge in general shows a similar understanding. He describes components of knowledge as a conceptual grasp of a theory, techniques we master to deal with practical problems and private perceptiveness needed to put such techniques to use in a variety of situations.

While conceptual, technical and practical knowledge have expanded at an exponential rate since industrialization, they have become increasingly insular. The humanities have been designated to the second order and ethics has been sidelined in our decision making. However, complementarity emerging across science, economics and law reflects intersections of knowledge based on common logic and ethics that are consistent with an Earth ethic that could guide us through the Anthropocene (Dibley, 2018).

I describe this response to the crisis of the Anthropocene as the re-emergence of common sense in our decision making. It offers an opportunity for the development of a coordinated approach to the challenges of the new epoch – an opportunity to seize a historic moment and conscientiously build an overriding synthesis of knowledge for our time. This chapter recognizes in the ancient concept of *sophia* the promise of new thinking in the Anthropocene.

Reinstating common sense

I describe thinking as being either passive or active. Passive thinking draws on knowledge that an individual uses subliminally – what Polanyi (1958) calls tacit knowledge, Midgley (2006) describes as 'habituated background to all our more detailed thinking' and Toulmin (2003) refers to as 'pre-theoretical life'. In the literature, this type of internalized knowledge has been found to include cultural beliefs, values, attitudes and mental models (Botha et al., 2008). Social and political theorist and philosopher Isiah Berlin (1998) argues that while people have different value sets, there are some values that we hold in common – a 'plurality of values which men can and do seek'. He goes on to say: 'The pursuit of (these values) is part of what it is to be human, and this is an objective given.' What then are these common values? What is the common experience? Where might we find them? In religious teachings, in the ideals of the French Revolution, perhaps linked to Maslow's (1943) *Hierarchy of Needs*, the United Nations *Universal Declaration of Human Rights* (1948) and even in school curricula?

A thorough investigation of these common values is beyond the scope of this chapter. However, something can be gleaned from the literature, more often in descriptions of what is missing from contemporary decision making; for example trust, experience, insight, judgement (Toulmin, 2003), truth (Skyttner, 2005) good citizenship and compassion (Sachs, 2011) and ethics and affirmation of life (Erikson, 2015). With such fundamental values in mind, scientists Brown and Harris (2014) assembled a set of common-sense questions (introspective, physical, social, aesthetic, ethical, sympathetic and reflective) that linked the subjective experience of individuals, the social purpose of a community and the pragmatism of an organization to the rationality of specialization. This internalized and habituated thinking shaped by maturation, experience and culture and common sense is the wellspring of trust, insight, good judgement, truth, citizenship, compassion, ethics and appreciation.

However, if common sense is passive thinking, operating at a tacit or habituated level, then there is a danger that it may be limited by self-interest. Something more is required to ensure our decisions support the best possible response to our dire environmental problems: the requirement that our decisions be reasonable. Reasonableness is expressed in active thinking and describes the way we engage in problem solving. This typically requires balancing potential solutions, including those that appear to satisfy self-interest, with the full range of effects including environmental impacts.

Toulmin (2003) describes reasonableness metaphorically as steering a middle way and keeping on an even level. He advocates adding rhetoric to our arguments whereby we can test the solidity of data supporting an argument and how far the resulting claims carry conviction. He maintains that reasoning procedures generally and argumentation must not merely consist of designated propositions divorced from the larger patterns of human life. He asserts that 'whether or not general theoretical abstractions can apply always and everywhere' or nowhere in particular as Tom Nagel (1974) points out, reasonableness requires us to take into account the substance of everyday experience. I regard reasonableness – taking into account the context in which a problem sits – to be a function of common sense.

Rational thinking

In contrast, rational thinking is described as a 'formal analytical process and fixed path of reasoning producing absolute facts that are quantifiable and that provide the basis for reliable knowledge' (Skyttner, 2005).

Historically, rational science and economics shared a common logic in the pursuit of consistency and stability, prediction and control, particularly in the role of equilibrium theory, in the history of economics and that of planetary stability in the history of physics (Toulmin, 2003). However, the pursuit of stability and certainty in both disciplines was shattered. The scientific world of physical determinism gave way to quantum mechanics and what we now call chaos and complexity theory and Keynesian and post-Keynesian economists declared general equilibrium theory to be neither accurate nor useful.

Regardless of contested histories, pre-eminent sciences, and later economics, have dominated the policy landscape with devastating environmental effect. By the 1950s the impact of the narrow application of rational science and economics was in full view. Gerald Weinberg (1975) describes the environmental impacts of applied science and technology as 'second-order effects produced by their first-order victories'. First-order victories might include those from medical science that have enabled us to live healthier, longer lives, from agricultural science that have resulted in increased food production and from technology and engineering that have led to large-scale manufacturing. Second-order effects are described by Weinberg as including environmental destruction and climatological changes, deforestation and desertification, acidification, decreasing subsoil water, shrinking ozone layer, decreasing biodiversity, extinction of species and the destruction of arable lands. I would add that climate change is further evidence of the folly of believing that applied science and technology bring stability and certainty. It reflects a failure to apply common sense to recognize that second-order effects or unforeseen consequences were bound to occur from major disturbance in an interdependent and interactive Earth system.

Similarly, rather than creating certainty and stability, the application of standard economic theory to public life and policy is widely acknowledged as contributing to the environmental crises that define the Anthropocene. Traditional economics does not account for environmental damage. The routine application of utility-based cost-benefit analysis to policy and regulatory instruments has been seen to favour economic utility over environmental protection (Kotchen, 2011). The narrow interpretation of utility (economic efficiency) applied by policy makers places the environmental effects of production and consumption (externalities) that would otherwise affect consumer utility and enterprise outside the market mechanism (Dibley, 2018).

The effects of the application of classical scientific and economic theory are illustrated in a case study of traditional rice farming in Bali. Following the advice of the Asian Development Bank to increase yield, the Indonesian government replaced traditional rice-farming methods with uncoordinated multiple cropping. This exposed native crops to insects, disease, drought and floods. Furthermore, it undermined the people's respect for the institutions that grew up around the waterways and practices, what Toulmin (2003) refers to as the 'moral infrastructure' of local society. Similarly, research undertaken by Ostrom (1990) showed that where authorities arbitrarily replaced age-old collective agreements of traditional farmers and fishers with a one-rule one-regime-fits-all approach to the use of common pool resources, the results were community conflict, environmental collapse and loss of livelihoods. What was missing in both cases was the application of common sense to account for the context and situation in which judgements about the application of science, economic theory and law were made.

The separation of common sense and rational knowledge has been the subject of much attention in the literature and I argue that common sense in the Anthropocene needs to be informed by science, economics and law (rational knowledge). However, while common sense can be informed by specialist (rational) knowledge on how things work, what can inform insight and judgement (what matters)?

The arts and humanities

Midgley (2006) reminds us that in other times, the poets and prose writers – Shakespeare, Pope, Blake, Tennyson, Nietzsche and Eliot – powerfully shaped the spirit of their respective ages. The essence of the arts and humanities is exploring, husbanding and debating values (Young, 2000) and our hopes and fears for the future. In our time, David Malouf (2015), in a collection of public writings on Australia, asserts that: 'Our uniqueness might lie ... in the tension between environment and culture rather than in what we can salvage by insisting on either the one or the other.' Malouf refers to the poetry of Kenneth Slessor (1901–1971) as recognizing what novelist Henry James (1843–1916) called our 'complex fate', which re-grounds us 'both in the particular world and in our own skin'. This is expressed in Slessor's poem, *South Country*, most poignantly in the final stanzas.

The Australian creative arts attest to our cultural connections with landscape – to be drawn on as we reground ourselves both in 'the particular world and in our own skin'. I conclude that while common sense is informed by institutional (rational) thinking, it is nourished by the arts and humanities (creative thinking) (Dibley, 2018).

An evolved common sense

When diffused with common sense, the full contribution of disciplinary and creative knowledge can be appreciated. A case study of regenerative farming demonstrates the full integration of ways of knowing. Charles Massy's original mechanical approach to farming in the 1980s reflected the application of the scientific and economic thinking of the day that failed to take into account Earth's natural systems. Massy also recognized a social or human factor in the application of theory to practice as human agency triggering landscape regeneration by working in harmony with natural systems. Undertaken conscientiously, Massy (2017), Marsh (2016) and other regenerative farmers apply common sense to agriculture. They include the application of scientific knowledge to their particular circumstances with regenerative and commercial effect. But there is something more. While regenerative farming places Massy squarely in the Earth ecosystem, it also represents a new archetype for our relationship to land that acknowledges not only what we are – members of a biophysical system – but also who we are – conscious and aware, capable of a morality that respects Earth. The farmer can be at the same time scientist, philosopher, poet and enabler – for Massy the return of the reed warbler is the 'heartfelt cry for all creation–a metaphor for us humans to once more become the enablers, the nurturers, the lovers of Earth' (p. 508). I regard this heightened awareness as the most evolved expression of common sense. It universalizes our experience and embraces our kinship with all beings. I have invoked the Greek word for wisdom, *sophia*, which Aristotle recognized as essential to our knowledge set, to describe the ultimate expression of common sense, the synthesis of all our knowledge.

Sophia is also present in the concept of cultural landscapes, which describes how landscape permeates our understanding of who and what we are. In 1992 UNESCO recognized three categories of cultural landscape with outstanding universal value for World Heritage listing. Recognition was predicated on the understanding that cultural landscapes are at the interface of culture and nature including tangible and intangible heritage, biological and cultural diversity. In addition, they are viewed as representing a woven net of relationships, which are the essence of culture and people's identity (Rossler, 2006). Within academia, cultural landscapes are not confined to the iconic. Rather, any system of interaction between human activity and natural habitat is regarded as a cultural landscape. In a sense, this understanding is broader than the definition applied within UNESCO, including, as it does, almost the whole of the Earth's occupied surface, plus almost all the uses, ecologies, interactions, practices, beliefs, concepts and traditions of people living within cultural landscapes (Fowler, 2003). Accelerated landscape degradation in the Anthropocene has an added poignancy when understood as the destruction of a cultural landscape. It is the shattering of our sense of identity and place. It sheds light on the cultural dislocation and loss resulting from the estrangement of indigenous people from their lands; and perhaps on the general decline in human self-confidence noted by Bookchin (1980). It explains Massy's relationship with the farm and the joyful significance of the return of the numinous reed warbler.

Sophia is present in emerging ethical and logical intersections in science, economics and law. Earth systems science places human beings in the interdependent Earth system, putting an end to science as a detached reflection of reality and recognizing it as a socially embedded practice. Ecological economics recognizes that protection of the environmental commons – the Earth system – is fundamental to the management of financial risk. Earth jurisprudence seeks to redefine the legal relationship between human and non-human entities and to develop laws that support the integrity and health of ecological systems. The application of a transdisciplinary ecological theory of value is demonstrated in regenerative farming.

Sophia in our time

Hamilton (2017) points to the irony of our current circumstances, arguing that only now, on the brink of self-annihilation, are we in a position to reach our full moral development – to ask the epoch-defining question – will we use our power to live harmoniously on Earth or to destroy it? He describes this defining choice as one between care and neglect. It goes to the heart of who we are.

Who shall we be in the Anthropocene?

The history of human beings is marked by our existential struggle to reconcile our functionality in a biophysical Earth system with our agency as a highly evolved species operating in our own constructed reality – our struggle to reconcile our vulnerability with our power. Contradiction is observable in the way we describe ourselves as being in relationship with nature (human beings in relationship with the rest of the Earth system, us and it) and in our perception of environmental degradation as separate from us.

Anthropocentric metaphors that have been used to describe our relationship with nature include nature as machine and nature as factory. They place human beings at the centre of the Earth and even the universe. They are rationalized by the doctrine of utilitarianism, which regards nature as an instrument for human ends without taking into reasonable account the needs and rights of other lifeforms and Earth systems. Recent policy metaphors such as the triple bottom line and ecosystem services have attempted to reconcile economic growth with environmental protections, while remaining committed to an anthropocentric world view that attributes instrumental value to nature.

The cultural separation of humans from nature is indefensible in the Anthropocene: Earth politics has changed. Our destiny as members of the Earth system (what we are) is inexorably tied; we are inseparable in geological time. But who we are has creative possibility. Will we use our agency to care for or to neglect Earth? As moral beings, what shall we decide is right or wrong? We find ourselves at a moral crossroads. In the Anthropocene, there is a new ontology that casts doubt on much of the thinking of the past on what it means to be human. Kant's (1795) *Categorical Imperative*, for instance, gave central importance to the dual nature of persons being expressed in respect for one another. It is a person's subjectivity, having a subjective point of view that has a special moral status according to Kant; every person must be regarded as having intrinsic value, while mere objects are simply a means to an end (Schoenecker, 2015). The moral challenge now is to reconcile the dual nature of persons as subjective conscious beings in a new objective reality, as members of the Earth system, the species *Homo sapiens*. Would Kant's *Categorical Imperative* recognize a new morality that understands a person's intrinsic value and the intrinsic value of nature as one? The defining reality of the Anthropocene calls for new thinking on what it is to be human.

Reconciling who and what we are with what we claim to know

I have found new thinking that could support an Earth ethic as our moral compass in the Anthropocene. I describe an Earth ethic as humanity's ethical and pragmatic commitment to express in all our thoughts and deeds, our indivisibility with nature. This places the observations of science in an interdependent Earth system that includes us, ensures the production, consumption and transfer of wealth operates within the limits of our finite planet and attributes legal status to the whole of the Earth system (Dibley, 2018).

In this new thinking, I recognize the unifying power of *sophia*, describing it in the aphorism: the bringing together of who and what we are with what we claim to know. 'What we are' refers to us as physical human beings, a functional part of an indivisible Earth system; 'who we are' refers to us as conscious beings, with agency and capable of making moral judgements about right and wrong; and 'what we claim to know' refers to us as the powerful makers and users of knowledge. I regard *sophia* as the reconciliation of biology, morality and agency – the bringing together of function, feeling and form. Attaining perhaps full moral development (Hamilton, 2017) – the epitome of knowing.

The synthesis of knowledge and experience that I call *sophia* is recognized elsewhere.

The collective mind

In 1955, archaeologist Teilhard de Chardin considered the next human evolutionary advance would be towards a collective brain, drawing on all human capabilities together as one, while still respecting the individual mind (Teilhard, 1955). In 2009, lawyer Christopher Stone argued that a specific type of human mentality and culture can, and sooner or later will, emerge (Stone, 2009). He called it the 'mind of humanity'. There are already expressions of this collectivity. Regenerative farming respects the indivisibility of all life and affirms our long history of coevolution with our landscapes and non-human animals. Massy describes this heightened awareness of our place in the scheme of things as an 'emergent mind' and 'the next stage in our cultural–biological evolution' (Massy 2017). Leopold (1949) considered the ability to see the cultural value of wilderness as an expression of 'intellectual humility'. And we have given the humanity–nature interface a name: a cultural landscape.

This knowledge conversion has been the focus of a large body of work undertaken over several decades by Valerie Brown, summarized in Brown (2008), Brown, Harris and Russell (2010) and Brown and Harris (2014). Her work shows that while the dominant mode of structuring knowledge in Western society is in separate compartments, marked by individual, community, expert, organizational and holistic ways of knowing, individuals are capable of breaking free from those compartments to access their own innovative thinking by engaging in a collective learning cycle.

Conclusion

In this enquiry, I found new thinking in the zeitgeist of the Anthropocene. Individuals are diffusing knowledge, applying an evolved form of common sense, *sophia*, which recognizes the humanity–nature interface as a cultural landscape, and connects common logic and ethics across existing silos of knowledge. Scientists, economists and lawyers are infusing institutional knowledge with common sense. Regenerative agriculture recognizes the place of the farm and the farmer in the Earth system. This emerging thinking may lead to a new ontology that is driven by a moral imperative to bring together who and what we are with what we claim to know. I conclude that *sophia* recognizes our undeniable place in nature, our common destiny with Earth's natural systems and our capacity to create new ways of thinking. At the conceptual level, this could uphold an Earth ethic. At a pragmatic level, it holds out the possibility of a reconstruction of knowledge – a new epistemology based on a convergence of thinking across existing knowledge cultures that accepts common sense and uncertainty.

Acknowledgements

I wish to thank the Fenner School of Environment and Society, Australian National University, for the opportunity to undertake my PhD and acknowledge the support of my PhD panel, Rob Dyball, Valerie Brown and David Freudenberger. I wish to acknowledge Val Brown as a wonderful mentor who has expanded my thinking and directed my reading beyond my original experience. I treasure our conversations over tea and sensible cake. I also wish to thank Anthony Hogan for encouraging me to undertake a PhD in the first place and for his ongoing support. I wish to acknowledge the ongoing support of the Fenner Writing Group whose individual and collective intellectual rigour and encouragement has been inspirational. I also wish to acknowledge the people from many disciplines with whom I have worked who see the big picture. We have recognized one another across rooms in meetings, negotiations and disputes. Finally, I wish to thank my partner Mark Thornton whose unwavering support, loving, wildly enthusiastic and intellectual, has been the mainstay of my academic endeavours.

References

Aristotle, (1953/2004) *The Nichomachean Ethics*, 384 /322, J. A. K. Thompson (tr.). London: Allen & Unwin.

Berlin, I. (1998) 'On Pluralism', *New York Review of Books*, XLV(8).

Bookchin, M. (1980) *Toward an Ecological Society*. Montreal: Black Rose Books.

Botha, A., Kourie D. and Snyman, R. (2008) *Coping with Continuous Change in the Business Environment, Knowledge Management and Knowledge Management Technology*. New York: Chandice.

Brown, V. A. (2008) *Leonardo's Vision, A Guide to Collective Thinking and Action*. Rotterdam: Sense.

Brown, V. A. and Harris, J. A. (2014) *The Human Capacity for Transformational Change: Harnessing the Collective Mind*. New York: Routledge.

Brown, V. A., Harris, J. A. and Russell, J. Y. (eds.) (2010) *Tackling Wicked Problems: Through the Transdisciplinary Imagination*. London: Earthscan.

Dibley, D. (Submitted 2018) *Sophia in the Anthropocene: An Earth Ethic for our Time*. PhD thesis, Fenner School of Environment and Society, Australian National University.

Erikson, J. D. (2015) 'The Unfinished Journey of Ecological Economics', Foreword in Brown, P. and Timmerman, P. (eds.), *Ecological Economics for the Anthropocene: An Emerging Paradigm*. New York: Columbia University Press.

Folke, C. S. R., Carpenter, B., Walker, M., Scheffer, T., Chapin and Rockström, J. (2010) 'Resilience Thinking: Integrating Resilience, Adaptability and Transformability', *Ecology and Society*, 15(4).

Fowler, P. (2003) *World Heritage Cultural Landscapes*. UNESCO World Heritage Centre, France.

Hamilton, C. (2017) *Defiant Earth: The Fate of Humans in the Anthropocene*. Crow's Nest: Allen & Unwin.

Kotchen, M. (2011) 'Cost Benefit Analysis', in Schneider, S. and Mastrandrea, M. (eds.), *Encyclopaedia of Climate and Weather*, 2nd ed. Oxford: Oxford University Press.

Leopold, A. (1949/1981) *A Sand Country Almanac and Sketches Here and There*. Oxford: Oxford University Press.

Malouf, D. (2015) *A First Place, Australia: Vintage Books*. New York: Random House.

Marsh, D. (2016) ecoag.org.au/a-blog-on-holism-from-david-marsh-boorowa/ [accessed 10/31/18].

Maslin, M. and Lewis, S. (2018) 'Anthropocene V Meghalayan: Why Geologists Are Fighting over Whether Humans are a Force of Nature', *The Conversation*, Melbourne, Australia.

Maslow, A. (1943) 'A Theory of Human Motivation', *Psychological Review*, 50(4).

Massy, C. (2017) *Call of the Reed Warbler: A New Agriculture – A New Earth*. St Lucia: University of Queensland Press.

Midgley, M. (2006) *Science and Poetry*. New York: Routledge.

Nagel, T. (1974) 'What Is it Like to be a Bat?', *Philosophical Review*, 83(4).

Ostrom, E. (1990) *Governing the Commons: The Evolution of Institutions for Collective Action*. Cambridge: Cambridge University Press.

Polanyi, M. (1958) *Personal Knowledge*. Chicago: University of Chicago Press.

Pugh, T. and Renner, M. (2014) *Failing Governance, Unsustainable Planet, Governing for Sustainability*. Washington, DC: Island Press.

Rossler, M. (2006) *World Heritage Cultural Landscapes*. London: Taylor & Francis.

Sachs, J. (2011) *The Price of Civilization, Economics and Ethics after the Fall*. New York: Random House.

Schoenecker, D. (2015) *Immanual Kant's Groundwork for the Metaphysics of Morals*. Cambridge, MA: Harvard University Press. See also www.philosophypages.com/hy/ 5i.htm [accessed 10/25/2018].

Skyttner, L. (2005) *General Systems Theory, Problems, Perspective, Practice*. Singapore: World Scientific Publishing.

Slessor, K. (1992) *South Country, Collected Poems*. North Ryde: Angus & Robertson.

Stone, C. (2009) *Should Trees Have Standing*? Oxford: Oxford University Press.

Teilhard de Chardin, P. (1955/1961) *The Phenomenon of Man*, Wall. B. (tr.). New York: Harper Torch Books.

Toulmin S. (2003) *Return to Reason*. Cambridge, MA: Harvard University Press.

Universal Declaration of Human Rights, adopted by General Assembly Resolution 217 A(III) of 10 December 1948.

Waters, C. N., Zalasiewicz, J., Summerhayes, C., Barnosky, A. D., Poirier, C., Galuszka, A. et al. (2016) 'The Anthropocene is Functionally and Stratigraphically Distinct from the Holocene', *Science*, 351, p. 6269.

Weinberg, G. (1975) *An Introduction to General Systems Thinking*. New York: Weinberg & Weinberg.

Young, R. M. (2000) 'Science and the Humanities in the Understanding of Human Nature'. Inaugural Lecture as Professor of Psychotherapy and Psychoanalytic Studies, Centre for Psychotherapeutic Studies, University of Sheffield, 25 May.

9

The organic, the mechanical and the emergent mind

Charles J. Massy

The emergence of the mechanical mind

In late 2014 a shocking story broke in the media. It concerned the murder of a NSW government environment compliance field officer in northern NSW while attempting to prevent unlawful vegetation clearing on potential cropping land.

The illegal clearing was in defiance of environmental laws that were seen as whittling away at farmer autonomy and freedom. Beliefs and feelings about the right to do what they wanted with freehold land had festered like corrosive acid in one family in particular near the town of Croppa Creek, especially in its 79-year-old patriarch, Ian Turnbull. For years he and his family had been engaged in long-running and bitter legal disputes with the NSW Department of Environment over its blatant pioneering approach to stripping virtually all vegetation off 'their' land.

As part of this legal battle, 51-year-old NSW environment compliance officer Glen Turner (who was generally liked and respected by farmers) had become the fall guy in handling the legal cases and liaison situations. But on 29 July 2014, matters boiled over for Turnbull. Near dusk, officer Glen Turner and a colleague were investigating further illegal clearing issues beside a dirt road adjoining the boundary of one of the Turnbull's properties. Suddenly

Turnbull pulled up alongside the officers' vehicle. As journalist Richard Guilliatt recounted, 'the old farmer' deliberately and systematically 'fired his rifle from a distance of 10–15m … a fifth shot hit Turner in the back, killing him. After the final shot, Turnbull told Turner's companion: "You can go now – I'll be home waiting for the police".' Turnbull was subsequently arrested for murder and, in mid-2016, was convicted. He will spend the rest of his life in gaol (Hoerr and Dole, 2016).

This case reveals the deeply ingrained mental constructs (the basic metaphorical, brick-like building blocks of our minds) that have become embedded in the subconscious of many post-1788 settler-descendants in this country. Only such powerful drivers could lead to the irrational explosion of passionate hate that ended in this murder on a lonely and dusty road in northern NSW. It includes a package of a number of deep attitudes and constructs that are part of what, as we will see directly, is the 'mechanical metaphor' of Western society. Foremost of these is that man is separate from, and so can dominate and is master over nature.

The question is: where did this 'mechanical mind' come from? It was this thinking and its mental constructs and attitudes that lay behind the extraordinary desecration of our Gondwanan ark and the longest lived culture in the world that began in 1788 and that continues still. We will therefore take a quick gallop across time, beginning with the age-old organic mind, followed by the emergence of a worldview like that of the mechanical mind (and more recent sub-components like economic rationalism and neoliberalism). On then to the emergent mind, which goes to the very heart of this book and the story of landscape and societal regeneration and human transformation.

Indigenous hunter-gatherers and nomadic herding people, who evolved in and live close to Mother Earth, exist in a different physical, mental and spiritual world to those of us in modern Western society. Their organic minds experience a reality that is an inter-connected living and spirit-filled world and landscape which holds that non-human entities: plants, animals, inanimate objects (like rocks, rivers, mountains) or phenomena (like thunder, wind, shadows), possess a living soul or spiritual essence and hence have awareness and feelings sometimes called *animism.* Australian indigenous peoples' traditional dances and stories describe the ways in which they are one with their environment.

By comparison, we wander around like blind, deaf and senseless people, oblivious to this other world. This is because modern Western society has arrived at a mental place that accepts the Cartesian dualism, which holds that our subjective and objective reality are two different things and which therefore separates our culture from nature. That's why modern Western humans treat the world as a detached mechanical object in narrow spheres of our activities.

Prior to the beginning of agriculture, the indigenous animistic view had reigned for 200,000 or more years in human societies. It constituted the 'organic' metaphor or 'mind' and its significance is that it contains strong ethical and spiritual implications for nurturing and sustaining the Earth: an associated value system.

For Australia's indigenous people, this organic world – where humans were a part of (and thought, felt, believed and acted as part of) nature – began to change after 1788. Consequently, so did the management of Australia's landscapes. This sudden and massive disruption to a unique Gondwanan ark and its ancient peoples was simply because white Europeans brought a totally different worldview and it was the building blocks of this worldview – the varied but connected and historically evolved 'rational' beliefs of modern Europe (and their constituent mental constructs) – that served as the powerful hammers that shattered both the indigenous world and Australia's eco-systems (Massy, 2017). Equally sobering, it is the same mind that has precipitated us into the potentially suicidal Anthropocene today. Therefore, if we are to turn around the Anthropocene along with the way we trash landscapes and the deeply prejudiced and racist way we treat others who seem different (such as our indigenous people), then we need to understand what comprises the deep psychic drivers of our behaviour.

The shift from an organic mind to our society's dominant mechanical worldview was triggered by the development of domesticated agriculture some 10,000 or so years ago. This comprised the first decisive shift in the metaphorical mind: when domesticated plants, animals and other natural phenomena had become manipulable 'property' and de-sacralized beings or entities. Consequently, from this time until the Renaissance of late 15th-century Europe, humans on the European and southwest Asian continents began a slow process of progressively throwing off the long coevolved organic mind or metaphor that had indivisibly bound them to nature, Mother Earth and a spiritual world.

This new cultural practice of agriculture and a greater abundance of food led to population growth and intensified urban living and to the rise of associated cultural practices, of the urban revolution and the first cities. Thence came the rise of large-scale political and social systems, all an increasingly far cry from our genetic evolutionary package as hunter-gatherers embedded in the organic mind.

Leading up to the Renaissance and Reformation there was a coalescence of powerful forces in Western culture that weakened the remnants of the organic metaphor and opened the way to an inculcation of the mechanistic metaphor. Evolving new city states and societies came to be held together by cooperation networks based on 'imagined orders', beliefs based on the shared imagination of thousands and then millions of people, on shared myths and stories, whether it be Hammurabi's code, the Egyptian gods Ra and Isis or the Christian one God.

The major phase of the metaphorical shift to the mechanical, argued Carolyn Merchant, really began between the 16th and 17th centuries when 'the image of an organic cosmos with a living female earth at its centre gave way to a mechanistic world view in which nature was reconstructed as dead and passive, to be dominated and controlled by humans'. The end result, in modern times, is that Western society no longer regards nature as a nurturing mother, holds her in reverence or treats her with respect. In a short time, as we will see, this led to the rise of modern capitalism (Merchant, 1980, pp. xvi, 125).

All the 'greats' of modern Western thought, science and philosophy played key roles in this shift. In the end, what ensured the entrenched dualism and removed objectivism of man versus nature was the partnering of science, philosophy, politics and religion during the critical phases of the scientific and commercial revolution of the Enlightenment that culminated in the 18th century.

Critical in this entire shift of the Western metaphor was the move from feudalism to capitalism. It was the machine image under Descartes and compatriots that invoked human (and especially male) power over nature – a forerunner philosophy to modern capitalism. Thus the Cartesian perception of nature as being dead, lifeless and mechanistic would prove the foundation stone of materialist reductionism, empiricism and objectivism (Descartes, [1637] 1955, pp. 1–119; Descartes [1637] 1952).

Then came the remaining piece in the intellectual construction of Enlightenment thought: the linkage of the mechanistic ideas of Bacon, Descartes, Newton and peers to that of economic and political theory so as to embed a capitalist philosophy. Such a philosophy embraced an intensified human instrumentalism in regard to nature, where nature was regarded as having no value except where it was reduced as a means to human ends, as an instrument for use. This bridge and a moral handwashing of humanity's exploitation of nature was made by John Locke and Adam Smith.

This, via John Locke's ideas, established a platform for the value system of the European Enlightenment: that through the use of reason (rationality) then one could profitably appropriate the whole sphere of nature as 'reason's own individual property' (Plumwood, 1993, pp. 111, 192). The idea of the acquisition of private property was the gateway to enabling humanity to 'progress' from the 'state of nature' into ordered civil society.

This shift to capitalism marked that moment when the traditional organic model of communal, organic interdependent society (one that emphasized the whole and not the parts) was undermined and transformed by competitiveness and acquisitiveness. Inherent now was an intellectual arrogance towards nature, where nature had become the raw material for wealth creation, with next to no moral/ethical restrictions on this.

It was only on analysing this entire story that it became clear to me why those early colonists and settlers who reached Australia were able to so unfeelingly plunder, destroy and kill and why these attitudes have deepened since with the rise of the 'culture of greed' and of economic rationalism. What also became clear is that this monumental shift in the collective human subconscious mind to a mechanical metaphor has huge relevance to the personal mental constructs – the attitudes and worldview – of modern Australian farmers and consequent land use (as so powerfully illustrated at the start of this chapter).

This is because what had fundamentally emerged was an entirely new imagined order of the Western mind that became the fundamental mind metaphor that drove much of Western behaviour from the Enlightenment onwards. Therefore, as Merchant rightly concluded, the mechanical metaphor not only constitutes 'the mainstream narrative of Western culture', but it comprises 'perhaps the most important mythology humans have developed to make sense of their relationship to the earth' (Merchant, 2004, p. 2).

This shift has awful complications and has led to a form of collective global amnesiac schizophrenia. That is, the modern blind belief in the tenets of economic rationalism, in limitless but necessary growth and a glorification of boundless greed means, according to John Ralston Saul, that the West in particular (but now spreading globally) 'has become addicted to a particular set of illusions in order to avoid coming to terms with its own reality' (Ralston Saul, 2005). This reality is that our unsustainable behaviour has precipitated us into the Anthropocene.

What is therefore clear is that not only did the modern 'rationalist' mechanical mind drive the post-1788 invasion and conquest of Australia, but that this mechanical mind metaphor and its dualisms reigns ever more powerfully still.

My recent book, *Call of the Reed Warbler: A New Agriculture – A New Earth*, concerns the emerging movement of regenerative agriculture. This involves an ecologically regenerating approach to agriculture, because it implies more than just sustaining something, but rather an active rebuilding of existing agro-ecological systems towards full health. Moreover, the process is ongoing in terms of improvement and positive transformation of landscape functions and thus landscapes themselves. This can involve the rebuilding or regeneration of soil itself and of biodiversity more widely, the recharging of aquifers, the production of healthier food, clean water and air and the replacement of external inputs along with the enhancement of social capital and ecological knowledge. Crucially, such a regenerative approach is built around the enablement of healthy self-organization of landscape functions (Massy, 2017, pp. 51–52).

In this book, the importance of 'mind' emerged as of central importance. This is because a new ecologically based agriculture contests the dominant industrial paradigm. So mind change is crucial to the regenerative process. Consequently, in the book, I attempted to describe a basic ecological literacy toolkit that allows farmers and others to get a simplified handle on how our landscapes work. Discussed are the solar, water, soil-mineral and ecosystem community dynamic landscape functions and how these are affected by the fifth function: that of the human–social (that is, the worldview or paradigm held by land-managers). What emerges is that none of these functions effectively acts in isolation, for all are indivisibly connected. And if we destabilize or simplify one then all the others are destabilized and in time degraded.

But what I discovered was that there is an even bigger picture: that these five landscape functions and cycles sit within a far larger matrix. This matrix comprises the macro operating influences, constraints and functions of nature in general. Some of these influences include factors like gravity and cosmic energy impacting Mother Earth. However, the basic operating features of our blue-green planet and its resources and of all life sit within a matrix of other wider processes and functional operating parameters.

The first and obvious macro factor is the constant operation of the mechanisms of natural selection. Part of this involves the ongoing processes of evolution, and thus coevolution. But in addition to these basic operating rules or mechanisms there is one other and over-arching crucial factor that plays a huge part in the story of regenerative agriculture, and this (as mentioned earlier) is the process of self-organization.

Self-organization is the propensity of not just living systems but other complex creative systems to regenerate and, seemingly spontaneously, provide solutions to triggered problems or destabilization. It comprises the inherent capacity of such systems (and especially natural systems), following disturbance or in the normal course of events, to move towards greater complexity and thus greater stability and resilience. This amazing capacity of complex creative systems, and thus of natural systems, has only begun to be articulated and understood in recent decades. The key to grasping the full significance of self-organization is that all healthy, functioning living systems comprise not only components that are all interconnected and interdependent, but also they are self-organizing. This is because their order and behaviour are not imposed by the environment but are established by the system itself.

In the case of the best regenerative farmers I have met, this landscape–human dialogue constitutes a true interactive dialogue; each appearing to be learning from the other. This, of course, in turn requires a great degree of openness and humility. Above all, the entire process is a creative one and so what is required is a different mind to that of reductionist or mechanical thinking. So it is important now (if we wish to appreciate how we actually exist, are embedded in, and operate within these wider complex creative systems) that we briefly acquaint ourselves with key elements of this fourth-dimensional matrix.

The new understandings about complex creative systems and things like emergent properties and self-organization have come out of a range of new thinking and research over recent decades: from systems thinking, chaos theory, complexity theory and new ecological, biological, physics, chemistry and other research. Great thinkers, philosophers and scientists of the past intuitively felt there were some deeper, inner drivers of life and earth's (even the universe's) systems: from Darwin, Wallace, Einstein to Popper, Ilya Prigogine and many more (Massy, 2017).

In his ground-breaking book *Cosmos*, begun in 1845, Alexander von Humboldt (the extraordinary polymath naturalist and empathic visionary and anticipator and influencer of Darwin) mused that the Earth was 'a natural whole animated and moved by inward forces' (Wulf, 2015). The intellectual giant Charles Darwin concluded his 1859 book *On the Origin of Species* with: 'From so simple a beginning endless forms most beautiful and most wonderful have been, and are being, evolved.' Nearly 70 years later in 1926 Jan Smuts, in his prescient book *Holism and Evolution*, also pondered on these big issues. 'Holism', he said, was 'the fundamental activity underlying and co-ordinating all others', thus revealing 'the universe as a Holistic Universe'. As Smuts expressed it, we live 'in a universe of whole-making' (Smuts, 1926/1973).

In 1977, along with Australian scientist John Eccles, philosopher Karl Popper was honest enough to admit that: 'The greatest riddle of cosmology is that the universe is, in a sense, creative'(Popper and Eccles, 1977). In the course of researching this area, it became clear to me that the processes of emergence and self-organization (each embedded in systems thinking) are the 'big engine', the big idea and reality that both underpins but also structures and organizes the functioning of our world. As such, we landscape- and macro-complex, creative systems-managers (i.e. we farmers, gardeners, national regional managers and such) need to be cognisant of, and simpatico with, this overarching functioning principle. That was what I discovered: That the leading regenerative farming innovators and practitioners most definitely are, even if they don't fully understand it.

Writer and philosopher on this issue J. J. Clarke, in his book *The Self-Creating Universe. The Making of a Worldview*, stated that there is a central theme which 'is the idea of creativity' and that this reveals 'how at all levels, from the cosmological to the human, the world demonstrates a universal inclination towards the emergence of new and unpredictable forms of order' (Clarke, 2013, p. 13). Moreover, the new knowledge and thinking surrounding emergence and self-organization is positive as it dispels the nihilism of a purely reductionist, mechanistic world and replaces it with an open-ended and positive view of creative evolution. As Clarke sums it: '*the new thinking … locates us in a world which is open and constantly transforming itself, creating new and astonishingly complex and beautiful forms*' (ibid.; my emphasis).

Nowhere more relevant to this story and humanity itself, therefore, is the idea that self-organization is a powerful property involved in the never ending development of the human mind and of consciousness itself. This book (as we have seen in previous chapters) is about this very process of transformation and transformation from the mechanical to the post-mechanical or what I now realize is an emergent mind. So, in every respect, but beginning at the most fundamental, that is, agriculture and its impact on the planet, human society and human health, this story becomes one of transformative change in our landscapes and selves via this giant, giant story of emergence, creativity and self-organization.

So, what is the core belief system and its parcel of mental constructs that differentiate regenerative farmers with an emergent mind from those in the dominant paradigm of industrial agriculture? From my extensive research over a number of years, I have noticed some major overall differences between traditional industrial farmers and those who have transformed to regenerative practitioners. Most startling (as revealed by action, belief and language) is the distinctive difference in the core building blocks of thinking – mental constructs – between industrial farmers (essentially of a mechanical mind) and regenerative farmers. These usually include a different set of ethics/values regarding land and nature. Moreover, transformative farmers are distinctively more biophilic or emergent in thought and language (especially use of metaphor). Associated with this is a deep emotional connection to their landscape.

They also appeared more holistic and more ecological in their thinking, while at the same time (in general) appearing more curious and hungry (or more open) for learning about other areas. This was commonly seen as an exciting, never ending journey worthy of considerable annual investments in time (up to a month per year) and money. In addition, they also exhibited great excitement and enthusiasm for their change in farming practices. And frequently women and husband–wife partners seemed to play bigger roles in the change process, in management and in partaking in the journey.

One of the most distinctive features was that a number of transformative farmers have taken the turn full circle back to incorporating old 'organic' components with new knowledge: confirming a conversion to an emergent mind. Another feature common to regenerative farmers is strong expressions of concerns and values linked to the importance of local community and of healthy food for healthy people. Practically, this is seen in involvement in farmers' markets, organic marketing and the development of their own food or fibre brands. This includes the cultivation of alliances to deliver healthy food from healthy landscapes direct to consumers so as to bypass the industrial food chain and big supermarkets.

The new emergent thinking is exemplified by one of Australia's leading and deepest thinkers in regenerative agriculture, David Marsh, who, with his wife Mary, farms near Boorowa in NSW. David's thinking has evolved through his own experience and also been influenced by a number of his many heroes, such as Aldo Leopold. It is no surprise, therefore, that, for Marsh, the way we farm and live has wider global philosophical implications. 'We are just fellow travellers with the other members of the biotic community', he says, and thus 'natural capital is the only true form of wealth' (Marsh, personal communication). To emphasize, he quotes another favourite author in Paul Hawken, who states that 'the economy is a wholly owned subsidiary of the environment' (Hawken, 1993). I argue with David that if we abdicate the thought process to someone else, then we can find ourselves playing the game others want us to play and getting on an economic treadmill that is hard to step off.

In a few sentences, David Marsh thus encapsulates not just the power of enabling all five landscape functions to indivisibly work in synergy, but a clear expression of an emergent mind: one open to ongoing disturbance, adjustment and equilibration. He thus calls his approach 'simplifying to complexity'. David encapsulates the essence of regenerative agriculture, while also throwing down an incontrovertible radical challenge to the self-annihilating agenda of the dominant ruling, growth-based consumptive paradigm of modern industrialized global society.

As described earlier, self-organization doesn't just involve a constant dialogue between the organism or landscape manager and its new environment or just with other critters and the energy and nutrients and matter flows in the system itself. No, in addition, this dialogue is wider still, with the Earth and its systems and ultimately the wider universe. Furthermore, the tools or mechanisms used for adaptation and response to environmental disturbance are novel traits that nature throws up: nature's emergent properties.

Self-organizing systems therefore are 'self-making'. This is the essential characteristic that distinguishes living from non-living systems. Self-making systems (whether they be ecological, other systems or organisms or even the human brain and Earth itself) therefore continuously recreate themselves within a boundary of their own making. This then leads to a recognition that such systems – and, in this case, humans and their society – are indivisibly connected to their environments (what leading thinkers in this field, Humberto Maturana and Francisco Varela (1992), call 'structural coupling'). What this means is that the internal structure and driving forces of self-organization within a system or organism are what drives the changes, adjustments and adaptation.

This entire process, therefore, is creative and positive in its response and drive towards the emergence of new yet unpredictable forms of order. To reiterate J. J. Clarke, this creative response of complex creative (adaptive) systems 'locates us in a world which is open and constantly transforming itself, creating new and astonishingly complex and beautiful forms' (Clarke, 2013, p. 13). This, therefore, is a total rebuttal of the nihilistic mechanism and meaningless of just chance and necessity. It is also a process that gives meaning, values and a spiritual dimension to our lives. As a consequence, I found in my investigations of regenerative farmers that enhanced meaning, high values and often spirituality were accompanying traits of their mind shift from the mechanical.

Therefore, part of an accompanying post-mechanical, post-ancient organic mind: the emergent mind, comprises a realization that we humans are indivisibly part of all life and Earth's living systems. The story of our own emergent evolution reveals this: that our 'descent' was and is not simply linear. Instead, increasing evidence leads to the conclusion that at each key evolutionary step (the vast amount of this in the microbial world, which dominates life's history), major progressive advances occurred through symbiotic mergers, mutualisms and cooperation of once different creatures. As Lynn Margulis and Dorian Sagan concluded: 'Life did not take over the globe by combat, but by networking. Life forms multiplied and complexified by co-opting others, not just by killing them' (Margulis and Sagan, 1986, p. 29).

So yes! We are related to plants, fungi, amoebae, newts, cockroaches, mosquitoes, dinosaurs and snakes, but also to whales and colourful songbirds, plus dogs, sheep, cattle and mountain gorillas. All are our cousins, and previous species our fathers and mothers. This alone, in conjunction with our structural coupling to our environment, surely behoves that the preservation of all life and of Earth and its systems should be our main ethical and moral concern.

We therefore face a giant conundrum. Through the long process of evolution and coevolution there arose the emergent property of the mechanical mind. We cannot put it back in a box because, like all emergent properties, it is nature's way of throwing up novelty: throwing up the experimental options for natural selection. Moreover, even if we could put such traits as our marvellous cognition back in a box, we would lose the good aspects: the capacity for problem resolution, creativity and imagination, the wonderful benefits in arts, science and innovation that have come both before, and especially since, the 16th century. Ironically, however, it is certain elements of these emergent properties of the mechanical mind (as described earlier) that have now mutated and become malignant.

So that is why a non-biophilic, amoral, economic rationalist global culture of greed and rapine is now the dominant, all-embracing and furthest reaching human cultural driver. The tragedy is that this is epitomized by modern industrial agriculture, the very same evolving human activity that led to humankind's cultural florescence. But now this agriculture flies in the face of the whole trend of life's evolution, which is built on networks, collaborations and symbiosis. Instead, this agriculture seeks to kill and dominate these arrangements, to grossly simplify and exert power.

Such a mechanical mind, therefore, is no longer truly emergent, given that the entire process of self-organization and its use of novelty and emergent properties is to re-equilibrate after disturbances so as to maintain an organism's or system's survival and enhance resilience. That is why it is clear that the mechanical mind has gone rogue. The challenge is that unless we as a species collectively act to constrain and negate the negative earth- and life-threatening aspects of the mechanical mind, then this once novel emergent property will be seen as the greatest of all failed experiments, remembering that life's history is full of extinct, ultimately ill-adapted species.

This is therefore the great challenge for humankind. Can we shift our modern industrial economic-rationalist, selfish, consumeristic, non-biophilic worldview to one that nurtures and not destroys? Can we shift to a truly earth-loving, ever open and adapting modern emergent mind?

I believe we can. This is because enormous crisis also brings opportunity and confrontation with the past–present that can stimulate change for the future. In Australian bush parlance: 'The thinking that got us into the shit won't get us out of it'; that having got ourselves into the Anthropocene only the best of our evolving modern brains and a totally new and different thinking will get us out of it. This can only come from a truly emergent mind.

This is revealed in the numerous examples in my book (and indeed this one), which provide a clear pathway to addressing and turning around the Anthropocene, through wide application of various forms of rapidly evolving agricultural practices and thinking. As Wes Jackson cogently observed: 'Since our break with nature came with agriculture, it seems fitting that the healing of culture begin with agriculture, fitting that agriculture take the lead' (Jackson, 1994, p. 26). In short, it is only when we begin to regenerate land and care for it – as opposed to overgrazing or degrading, destroying and intensively farming with monocultures and poisons – that our minds can shift to where we can begin to evolve a human–land ethic derived from and within place.

This lead by agriculture is not just occurring in Australia but across every continent. Moreover, this, in turn, is being linked to similar post-mechanical, emergent thinking surrounding healthy food and meaningful design and living in urban centres and in movements like the drive for Fair Food; and in organizations such as *La Via Campesina* – the global peasant-driven revolt against the mechanical mind and its dominant globalized, degrading powers.

Crucially also, the new regenerative agriculture is not tiny stuff on small backyard plots or half-acre allotments (although these, too, are part of the solution and some of the practices apply and are being practised there also). No, the practices involved are being applied on tens of millions of hectares, while others have the potential to be applied equally broadly worldwide: with new cropping approaches like grassland grains, pasture cropping and natural intelligence agriculture, or with agroforestry, farmer-managed natural regeneration, biological agriculture and many others (Massy, 2017).

Therefore, this revolution has the potential to make massive inroads into addressing the key Anthropocene challenges of climate change, biodiversity loss, correction of distortions to the nitrogen, phosphorous and other key elemental cycles, addressing land degradation, regenerating the water cycle and so on. In addition, this new regenerative agriculture has astounding potential to address human ill health, to therefore address the huge burden of modern health costs on society, to make farmers more profitable and free of the clutches of banks and large multinational chemical and energy suppliers, to therefore enhance both rural and urban societies at their interface, and to free farmers and others ensnared by the merciless behemoths of agri-business, commodity trading, retailing, food processing and the like.

In short, a widespread shift to a new regenerative agriculture has the potential to fundamentally change for the better modern industrial capitalist society. This is because it is a bottom-up movement and involves an entirely new way of thinking and feeling about farming, society and how we treat Mother Earth. As a consequence, we need to be aware that a new regenerative agriculture is subversive, but in a truly constructive way. That is why I call it an 'underground insurgency', for it is a bottom-up revolution instigated by a new mind: an emergent mind.

Acknowledgements

I believe it is uncommon to have a chapter in a book where one's fellow authors have all played a role in influencing, by various means, the thinking behind the chapter. But that is the case in this instance and I sincerely thank this collective brains trust. Behind this is an exceptional mentor in Professor Val Brown, ably assisted by Dr John Harris – and my heartfelt thanks go to them for this initiative. Val Brown was also instrumental in initiating the Fenner Writing Group – whose members, embracing holistic thinking and human ecology, adorn these pages. Thanks are also due to the Human Ecology Programme, currently at the Australian National University's Fenner School of Environment and Society. This programme was the first course to teach holistic teaching in Australia from the early 1970s and its fruits can be seen in these pages.

References

Clarke, J. J. (2013) *The Self-Creating Universe: The Making of a Worldview*. Bloomington, IN: Xlibris.

Descartes, R. (1637/1952) *Descartes' Philosophical Writings*, N. K. Smith (tr.), London: Macmillan. [Cited in Plumwood 1993, p. 114.]

Descartes, R. (1637/1955) 'Discourse on Method', in E. S. Haldane and C. R. T. Ross (eds.), *Philosophical Works of Descartes*, 2 Vols. New York: Dover.

Hawken, P. (1993) *The Ecology of Commerce: A Declaration of Sustainable*. San Francisco: HarperCollins.

Hoerr, K. and Dole, N. (2016) www.abc.net.au/news/2016-05-27/nsw-farmer-found-guilty-of-murdering-environment-officer/7452728 [accessed 10/17/2018]

Jackson, W. (1980) *New Roots for Agriculture*. Lincoln, NA: University of Nebraska Press.

Jackson, W. (1994) *Becoming Native to This Place. The Blazer Lectures for 1991*. Lexington, KT: University Press of Kentucky.

Massy, C. (2017) *Call of the Reed Warbler: A New Agriculture – A New Earth*. St Lucia: University of Queensland Press.

Margulis, L. and Sagan, D. (1986) *Microcosmos: Four Billion Years of Evolution from Our Microbial Ancestors*. Berkeley, CA: University of California Press.

Margulis, L. and Sagan D. (1995) *What is Life*. Rockefeller Centre, NY: Simon & Schuster.

Maturana, H. R. and Varela, F. J. (1992) *The Tree of Knowledge. The Biological Roots of Human Understanding*. Boston, MA: Shambhala.

Merchant, C. (1980/83) *The Death of Nature: Women, Ecology, and the Scientific Revolution*. New York: HarperCollins.

Merchant, C. (2004) *Reinventing Eden. The Fate of Nature in Western Culture*. New York, Routledge.

Plumwood, V. (1993) *Feminism and the Mastery of Nature*. London, Routledge.

Popper, K. R. and Eccles, J. (1977) *The Self and its Brain*. Berlin and London: Springer-Verlag.

Ralston Saul, J. (2005) *The Collapse of Globalism: And the Re-invention of the World*. Woodstock, NY: Overlook Press.

Smuts, J. (1926/1973) *Holism and Evolution*. Westport, CT: Greenwood Press.

Wulf, A. (2015) *The Invention of Nature: Alexander von Humboldt's New World*. New York: Vintage Books.

10

Escaping the 'circular conundrum'

Cropping and learning in Northern Australia

Kathryn E. Andrews

Background

Northern Australia has thwarted grand agricultural dreams and the global pattern of colonization. Yet ever since the arrival of Europeans over 200 years ago these dreams and the drive to achieve them continues.

Cropping plays a minor role in Northern Australia and an even less of one in the wet-dry tropics, barely contributing to economic production. Less than one per cent is intensively farmed. Since colonization, however, broad-acre cropping appears regularly in the political and public discourse about northern development, often in an optimistic light. The north has been feted as the next 'food bowl' or 'rice bowl for Asia'. And, with increasing international food security concerns and the clamour for development in Australia, there is another revival in public discussion of the great 'untapped' agricultural potential of northern Australia.

This ongoing cycle creates a 'circular conundrum' – a cycle of publicly expressed expectation for conventional broad-acre cropping in northern Australia, the failure of this to occur at the scale envisaged and then back to the high expectations. What drives this momentum and why does it appear to continue with little regard to outcomes or history or place? Why does learning not appear to occur? It seems to defy logic and piques curiosity. This chapter will briefly summarize the drivers of the 'circular conundrum', the cultural factors that hinder our learning and explore how this might help us understand relationship with place and how we learn in place.

In his renowned and controversial book of 1965, *The Northern Myth*, agricultural economist Bruce Davidson denies the potential for agriculture in the north and decries the wasted investment and expectations, contending that the economic case did not justify investment. My grandfather, C. S. Christian, a senior agricultural scientist who worked to establish agriculture across the north, owned a copy of the book that now sits in my bookcase. Under the title 'The Northern Myth' on the front page he has scrawled 'Or The dead hand of the economist', epitomizing the ongoing debate between believers and naysayers of northern Australian agriculture.

Failed attempts to deliver the mythical food bowl have accumulated; with millions of American dollars lost by ambitious Texan entrepreneurs and expended by Australian governments, Australian families walking off unviable farms with nothing and Indigenous Australians being even further marginalized. Scattered across northern Australia are fascinating and diverse stories of cropping, from stoicism and hardship to greed and profligacy, from individual backbreaking endeavours to presumptuous multinational companies and governments.

Modern farming is perceived to have advanced its frontiers and intensified production (Pretty, 2007), a trend that feeds into our expectation that intensive agriculture will, and should, occur across northern Australia. The success of Australia's agriculture through adaptation and innovation (Henzell, 2007) contributes to expectations of the North's productivity; if southern Australia can farm some of the driest and poorest soils in the world why not the North? Australia is dominated by the idea of industrialized export agriculture. As Powell says geographical transformation 'cannot be understood without examining the gap between rhetoric and reality' (1977, p. 83). The public discourse about northern Australia reveals our 'systems of knowledge and belief' (Fairclough, 1992, p. 36). Dryzek (2005, p. 9) reinforces this understanding writing: 'Discourses construct meaning and relationships, helping to define common sense and legitimate knowledge.'

The most commonly used boundary for Northern Australia, and most extensive, is everything above the Tropic of Capricorn: a line across a remarkably diverse landscape. The boundary I use refers to the area of the wet-dry tropics or tropical savannas of Northern Australia. This excludes the very different wet tropics of northeast Queensland (the only region in the north with significant cropping), although this still leaves an area larger than 1.5 million km^2 (about six times the size of the United Kingdom). The climate is distinctly different from that of the rest of Australia. Its bipolar seasonality, a wet and a dry, literally takes the country from flood to drought every year. As an old hand described, it is not just a wet/dry but a '*dry* dry and *wet* wet'.

Land use in northern Australia remains dominated by pastoralism, while economic production is led by mining and petroleum. Fifty per cent is now Aboriginal lands, and only 4.8 per cent of the Australian population (BITRE, 2011) lives across the top third of Australia and about 0.97 per cent in the northwest portion (Western Australia and Northern Territory), less than one person per 2.5 km^2.

How we define failure or success in the context of cropping in the north depends on many factors, revealing the continuing complexity of the debate; for some success cannot be determined by economic factors alone or the timeframes for success may be long term. Geographical scale is another crucial factor.

Constant colonizing

More than 60,000 years of habitation brings a deep and detailed knowledge of country. It brings a different relationship with country; a relationship embedded in an understanding of one's own place and role in the whole – family, community and environment. This world view is one that non-Indigenous Australians may be lucky enough to fleetingly taste, from the breeze, the fresh caught fish, the stormy horizon or the words of an Aboriginal friend. It is rare for anyone not born into it to truly comprehend it.

However, 230 years of colonization have scarred this knowledge of country at best and, at worst, in some places wounded it fatally. The occupation of Northern Australia was unrelenting and, in many places, brutal (Attwood and Foster, 2003; Rose, 2004). 'It was notorious', reported the Government Resident of the Northern Territory in 1900, 'that the blackfellows were shot down like crows, and that no notice was taken' (quoted in Farwell, 1951, p. 150). As with the rest of Australia there was and is 'an inherent conflict between the goal of colonising a new country and the rights of indigenous people' (Attwood and Foster, 2003, p. 6). The history of strong resistance to white invasion of the Kimberley is captured in the history of Aboriginal resistance fighter Jandamarra (Pedersen and Woorunmurra, 1995; Kinnane, 2008).

The British did not recognize Indigenous Australians as having any ownership of the country – the infamous *Terra Nullius* (land without owners) leading to policies and laws based on blindness or a lie built on ongoing denial of rights. It was empty country; perpetuated also by the belief that legitimate ownership is marked by certain types of use, cropping being one of them. Peoples perceived to be without agriculture were not truly in possession of law and rights and land. Recent work, however, has shown how starkly Indigenous land practices and management, including evidence of agriculture and settlements, were expunged from our history, an inconvenient truth that countered the necessary myth of *Terra Nullius* (Gammage, 2011; Pascoe, 2014).

Cropping stories

Exploring cropping examples over the last 200 years reveals astounding unbounded stoicism, naïvety and arrogance. When studying narratives from the early 1800s to the 21st century the drivers are visible, as are the variables that lead to failure – across scales from individual capacity to national policy to international markets. Additionally, narratives reveal emerging and persistent themes, including race, learning and relationship with place.

There was the fiasco of sugar plantations attempted in the Northern Territory in the 1880s with the name Delissaville still marking the site of a failed plantation near Darwin. Then the awkward life of the sugar industry in the Ord River region in Western Australia from 1994 to 2007, finishing with a Western Australian government bailout for local farmers and the South Korean company selling its sugar mill for as good as nothing. Factors included a lack of guaranteed supply for the mill, scale of production, global price and volatility, poor relationship between mill and growers and sugar quality.

Rice provides grist for the mill in both the Northern Territory and Western Australia; from the Chinese growers of the late 1880s to government researchers and Hollywood millionaires in the 1950s. This last was Territory Rice Ltd, a company funded by Hollywood naïvety and encouraged by Australian politicians. The company aimed to develop 200,000 hectares southeast of Darwin on the plains of the Adelaide River. At its peak they sowed just 2000 hectares.

In the 1950s with the threat of communism in southeast Asia the perceived 'yellow peril' that had driven the push to populate northern Australia also became red. The release of classified Cabinet papers revealed that in 1953 the communist threat in southeast Asia was used by the British government to convince the Australian government to help maintain rice supplies to British colonies.

Cotton in the north came to public prominence with the Ord Irrigation Scheme in the 1960s and even more so with the industry's dramatic failure in the 1970s when the crop was abandoned. By 1974 insecticide spraying had become the largest single cost for farmers. From 1964 to 1974, 435 tonnes of DDT and 412 tonnes of toxaphene were applied to the cotton crops. As someone said who lived and farmed there at the time: 'Brilliant place to live. There were no flies because we'd poisoned everything. DDT, Andrin, Parathene, you name it we used it.'

Peanuts are the stoic but less sexy cropping cousin of the others with a long history of peanut growing, including along the Daly River of the Northern Territory in the 1930s and 1940s. In C. S. Christian's unpublished journal of the 1946 CSIRO land survey, undertaken in a post-war government push to develop agriculture and populate the north, he describes an almost Dickensian scene of peanut growers on the Daly: 'The picture of these peanut farmers reminds me of the poor whites of S[outh] Kentucky.' And they relied on local unpaid Aboriginal labour. More recent attempts in the late 1990s and in Katherine in the early 2000s, by hard-working farmers and the well-informed Peanut Company of Australia, came to nought, with the Peanut Company having difficulty selling off its properties in 2012. Rainfall variability and unpredictability, temperature, diseases, the lack of a profitable rotation crop, high input costs, low gross margins – the list becomes a familiar one.

If it exists at all, the safe operating envelope for broadacre cropping is much smaller in the wet-dry tropics than in southern Australia. Less room for forgiveness, higher risk and lack of capital of all types creates greater vulnerability and threatens viability. The many cropping stories reveal the interwoven characteristics and relationships of the circular conundrum. What becomes visible is that the overarching narrative is not actually of agriculture and cropping, it is of peoples' relationships with place, and knowing place.

Drivers of cropping development

The impetus for agricultural development of northern Australia continues in the face of failure and obstacles. What are the ultimate drivers – those more stable and relevant to long-term policy making?

Analysing northern agricultural development since the 1820s using primary and secondary material, including contemporary conversations, I identified ten drivers for agricultural development: securing the nation; legitimating the nation; markets and trade routes; land of opportunity; making it white; feeding the world; plugging the money sink; Indigenous opportunity; climate change and the water crisis; and, finally, managing risk and diversifying. These drivers represent ideas that reflect the dominant culture and values of Australia over time; the power of ideas. Each of the ten drivers is, or was, relatively overt, as evidenced by appearances in the literature and media. (Andrews, 2015)

Analysis reveals that the drivers are removed from the environment itself, needing only general representations to be primed, rather than the real thing – cues such as space, apparent emptiness and water. They are the signifiers that carry the concepts. As the drivers are not actions themselves, but the expression of motive, they therefore stand separate from place. The drivers themselves do not necessarily emanate from the north; neither do they interact with the north and thereby don't have relationship with the environment. Yet together the signifiers and concepts create the myth itself, in this instance, the vision of agricultural development in northern Australia.

For those who are living and working in the north this is different. They are in relationship with place, directly influencing and influenced by it. And their relationship with the drivers can be tangible – a desire to make a go of their farm for their family or to create employment opportunities for their communities.

Hindered learning

> That's why I'm quite surprised because the old people would say that you see the land being buggered up over and over again. You can sit and watch it. One will come and bugger it up then go away, another will come and do the same thing over again. And the old people say that there are no lessons learned and they can't understand why they keep doing this 'cause it's a failed crop. (Mona Liddy, Wagiman elder Daly River Northern Territory, personal communication)

Accusations persist that Australia's colonizers have not been learning in the north; not from history, from experience or from the land itself (Price, 1933; Taylor, 1955; Bauer, 1963; Powell, 1988; Robin, 2007). This was to be expected in the early days of colonization when our knowledge 'was essentially a rich amalgam of fact, fiction and rampant guesswork (Powell, 1988, p. xiv). Yet some believe this remains an endemic problem in the north 'where history, particularly the history of failures, often goes unacknowledged' (Robin, 2007, p. 123).

Exploring the conundrum shows that learning is hindered by five broad factors: a patchy, lumpy and leaky knowledge system; skewed communication and manipulation of facts; inadequate governance and politicized decision making; culture, beliefs and northern myths; and scale, complexity and variability (Andrews, 2015). Through the many stories of attempted cropping we find that learning about Australia's North is not a linear, cumulative process. Rather learning, like the soil of the north, is patchy, occurring in different places, at different times and at different scales. Not only does learning occur in fits and starts in both research and practice, that is, it is 'lumpy', it is also lost in a variety of ways – it forms a leaky knowledge system. Where learning is occurring, it is mostly at an individual and community level, not a policy or societal level.

Culture is the most fundamental aspect of what encourages or hinders learning, influencing other factors such as how we define, create and share knowledge and what is communicated and why. It influences how we are governed and the decisions that are made. Threading through the northern narratives and discourse are powerful and persistent cultural themes, perhaps myths. Just as these dominant cultural beliefs hinder learning, they can also inhibit or disrupt learning entirely, perhaps even assist in forgetting. Learning is as much about the questions we ask as the answers we find to them.

The frontier mentality

> No longer will Northern Australia be seen as the last frontier: it is in fact, the next frontier. (Liberal Party of Australia, 2013)

Understanding the overt and ever present frontier mentality helps explain the circular conundrum and hindered capacity to learn.

The imposition of the desires of the centre on the periphery, as recognized in geographic literature, is not unique to Australia. Rather we are another example of internal colonization of the north from the south (Sörlin, 2002). Colonizing is not a thing of the past (Head, 2000). The 'circular conundrum' is accompanied by, even reliant on, a set of beliefs that together create the frontier. The northern Australian frontier relies on three beliefs, each of which inhibits learning. The first, an 'empty' North (Blaut, 1993), originates particularly from outside the north and brings with it blindness to Indigenous Australians' land ownership and roles, such as the view that they are a barrier to development. This is a critique made by Northern Indigenous leaders. Second, the frontier mentality can create an unwillingness to learn from what has come before; there is no history to learn from in this continual frontier. It is tabula rasa or the 'year zero' (Rose, 2004). Third is an unshakeable belief in the inevitability of a westernized development trajectory (Cronon 1987; Pretty, 2002; Sluyter, 2002), as discussed below. These beliefs are not an Australian anomaly rather, they are an expression of colonization.

Additionally, two knowledge disjunctions occurring with colonization are integral to the frontier mentality; ignoring Indigenous peoples' relationship with and expertise of country and the imposition of knowledge from entirely different environmental contexts.

Arrogance of the contemporary

> We are at crossroads in the development and establishment of agricultural endeavour (Everingham, 1982, p. xii)

Everingham, first Chief Minister of Northern Territory self-government, epitomizes this persistent belief of arriving at a turning point. Regularly in the discourse is a sense that this time will be different, whether an overt statement or implicit optimism. This may be justified by technological improvements, new government arrangements, better infrastructure, bigger markets, size of a venture or better agronomic knowledge. And who is to know that the optimism is misplaced? It is only with looking back through time that one sees a pattern. It does, however, create the impression of 'an arrogance of the contemporary'; we know better now and past attempts don't count. This belief carries a sense that previous barriers to cropping can now be conquered, with technology, knowledge and will.

Southern assumptions

Viewing the north through a southern lens can bring southern assumptions and expectations and impose an inappropriate southern agenda. Southern arrogance also inhibits learning. Just as European agricultural knowledge was once applied to southern Australia, southern knowledge comes with an assumption of its superiority. Examples abound of applying practices from elsewhere and making the same mistakes unnecessarily. Perhaps learning in a foreign environment depends on *un*learning.

Great expectations

Two fundamental beliefs, the development imperative and technological determinism, underlie the many drivers of the circular conundrum. They also impede learning. Technological determinism sees nature's problems as manageable through technology (Pretty, 2002), just as rice growing south of Darwin in the 1950s was touted by politicians as simply an engineering problem. The frontier myth is also tied to a belief in the inevitability of the Western development trajectory. For example, plentiful water and space must lead to cropping and productivity and wealth, as described in Ernestine Hill's *Water into Gold* (1937). To have this belief regularly frustrated through the failure of the accepted trajectory of progress affronts deep cultural tropes.

The result of this model of advancement is that the plurality of alternatives is not seen, alternatives to 'orthodox development ... based on intimate knowledge of the changing, complex realities of particular places' (Sluyter, 2002, p. 5). The ongoing push for conventional development prevents us from seeing alternatives outside this trajectory. With this perspective failure leads to a response that we just need more of the same whether dollars, land, people, political will or research. Canada's northern frontier shares similar 'northern visions' of a treasure trove to be exploited with the establishment of hard infrastructure such as railways and road building, and grand-scale plans. In Canada, also, '"Northern development" was virtually a political mantra' and seen as inextricably entwined with the future of the country (Wynn, 2007, p. 31).

From learning landscape literacy to complex systems literacy

The slow journey by colonizing Australians to cultivate landscape literacy in northern Australia relies on cultural landscape literacy and climate literacy and dependent on all three is a complex systems literacy. But this has grown slowly in the nonlinear nature of learning, with the fits and starts and backward jumps found in all close history.

Understanding the northern Australian landscape requires recognition that it is a cultural landscape; a result of tens of thousands of years of culture and law, with intricate, intimate land and sea management. We see a stuttering, but significant, progression from *Terra nullius* to recognition and revival of Indigenous Land Management and participation in the northern development agenda. There is a tension in Australian literature for white Australia to grow past their European beginnings and learn from the first Australians (Day, 2005); a struggle that is more than between knowledge domains.

Recognition of a cultural landscape is elemental for the health of northern Australian landscapes and a healthy future for its people. Recognizing that this place is the result of a long relationship with people brings three things: acknowledgement of existing Indigenous knowledge of country and respect for this knowledge domain; better understanding of the country; and the capacity to recognize and create multiple and varied opportunities for the future. A challenging parallel exists between the colonizers' original motive for cropping as a means to populate the north and current Indigenous goals of getting Indigenous people back onto their country. 'Bringing people back' and 'looking after country' is seen as fundamental to the health of people and health of country (Hill et al., 2013). To achieve this, we all need to shed the frontier mentality that haunts the north.

With the greater extremes and variability expected with climate change, climate and landscape literacy in the wet-dry tropics become even more vital.

Learning within a wicked problem?

The circular conundrum carries a heavy load: unrealistic though overtly rational expectations for cropping; the burden of constant colonizing – the forever frontier; and the powerful imperative to develop. Positive examples such as collaborative learning and inclusive processes exist but are not sustained or sufficient to tip into a changed system or approach. Several authors optimistically assume these are indicative of change (Kinnane, 2002; Holmes, 2010). However, it seems that we revert to the status quo.

Knowledge, scale and time in land management are captured by Stafford-Smith in relation to the Australian rangelands (Stafford-Smith and Cribb, 2009). He elucidates how a lifetime of understanding is no longer enough; the long-term nature of desert climate cycles with its slow responses mean 'that *individual* experience is quite inadequate'. Perhaps it has never been enough in Australia.

More than the ability to pass on individual knowledge between people, the crux of the learning challenge we face is stepping up from individual learning to societal learning. Social learning and collective thinking have been posited as ways for societies to develop ecological literacy, understand complex systems and respond to wicked problems (Keen et al., 2005; Henry, 2009; Brown et al., 2010; Swartling et al., 2010). This is done not through learning different things, but through learning differently.

I identify three levels of question and correspondingly three levels of learning. The first, quite constrained, sits within a specific cropping attempt responding to problems that arise; it is direct and applied – how to make this cropping scheme work? The next level up addresses cropping (and failed attempts) more broadly: how do we make this cropping industry or cropping in northern Australia successful? It is still bounded by the problem/solution approach. Finally, there is the meta level that encompasses the conundrum as a whole. It seeks to understand the system as a whole and to understand learning. At this level, the problem definition is opened up and learning becomes a redefining of the problem. It ceases to be establishing broad-scale cropping in northern Australia and becomes a redefinition of productive land use that is meeting the needs of the people and the place.

Power of place, relationship with place

These stories coalesce into one story. What we learn is that the overarching narrative is not one of cropping, rather it is one of peoples' and society's relationships with place.

While Australia's sense of the North remains only a representation – the agricultural ideal of empty land and plentiful water – it can carry cultural tropes such as 'the frontier' and harbour great expectations. Public and political discourse, particularly from southern Australia, regularly reveals this construction of the North. This defines and constrains the knowledge and approach that can be used to understand it. When we move from high expectations to cropping attempts, we are moving from ideas to place, from thoughts and aspirations to contextualized action; and that is when the North becomes an agent, the most powerful in all these narratives. We see played out through time the stark distinction between perception of place as opposed to relationship with place. In contrast to such a distinction, Rose (2002, p. 320) explains that 'the genius of Aboriginal Australians finds its greatest expression in a theory and practice of place'; an as yet unrealized praxis for settler societies.

There is a deep contextualized learning in a place. One size doesn't fit all, whether policy, investment, farming practices or research and development. The naïvety in approaching the north as a homogenous entity reveals the need for contextualized and local knowledge, identified by Christian (1959) who recognized long ago we need an iterative process with evermore detailed knowledge developed in a local context.

Alternative pathways have been travelled while the conundrum has been cycling. There are emerging opportunities and existing initiatives in cultural and ecotourism, carbon abatement and sequestration through fire management, biosecurity and conservation management and a significant rise in Indigenous Land Management organizations and jobs across northern Australia (Armstrong et al., n.d.; Yibarbuk et al., 2001; Hill et al., 2005, 2008; Russell-Smith et al., 2009; Holmes, 2010; Hill et al., 2013).

Many years and multiple initiatives have been building these opportunities locally and regionally. There is an evolution within northern Australia of our understanding of managing land systems for production and a redefining of agriculture and development. Is northern Australia moving on from constant colonizing? These alternative pathways are vulnerable; most are still reliant upon government, and the south. They are therefore still vulnerable to the whim of the political and policy context and the drivers above.

When viewed alongside the circular conundrum, however, these pathways offer a way through this wicked problem or at least a way to learn. A new way embedded in the old and embedded in place, yet all the while adopting and adapting science and technology, and the market. Pervading any discussion of the future is the enduring sense of loss entwined with stoic hope that Indigenous people have for country, conjured by Kinnane (2002) in describing the many visions, complex and contested, layered over his grandmother's country, now flooded and under Lake Argyle, for the Ord River Irrigation Area.

Acknowledgement

I recognize and pay respects to all Traditional Owners and custodians of Northern Australia past and present. My thanks for peoples' stories and generosity in the north, and a world away in Canberra - our writing group.

References

Andrews, Kathryn E. (2015) '"A Circular Conundrum" 150 Years of Cropping and Complexity in North-West Australia'. Unpublished PhD thesis, Australian National University.

Armstrong, R., Morrison, J. and Yu, P. (n.d.) *Indigenous Land and Sea Management and Sustainable Business Development in Northern Australia*. Darwin: NAILSMA.

Attwood, B. and Foster, S. G. (eds.) (2003) *Frontier Conflict – The Australian Experience*. Canberra: National Museum of Australia.

Bauer, F. H. (1963) 'Significant Factors in the White Settlement of Northern Australia', *Australian Geographical Studies*, 1, pp. 39–48.

Bauer, F. H. (ed.) (1977) *Cropping in Northern Australia: Anatomy of Success and Failure*. Darwin: ANU Press.

Blaut, J. M. (1993) *The Colonizer's Model of the World – Geographical Diffusionism and Eurocentric History*. New York: Guilford Press.

Brown, V. A. (2008) *Leonardo's Vision: A Guide to Collective Thinking and Action*. Rotterdam: Sense Publishers.

Brown, V. A., Harris, J. A. and Russell, J. Y. (eds.) (2010) *Tackling Wicked Problems through the Transdisciplinary Imagination*. London: Earthscan.

Bureau of Infrastructure Transport and Regional Economics (BITRE) (2011) 'Northern Australia Statistical Compendium 2011 Update'. Canberra: Australian Government.

Christian, C. S. [Unpublished] (1946) 'Northern Australia Survey Journal of C.S. Christian'. Personal records.

Christian, C. S. (1959) 'The Eco-complex in its Importance for Agricultural Assessment', *Biogeography and Ecology in Australia*. Canberra: CSIRO.

Cronon, W. (1987) 'Revisiting the Vanishing Frontier: The Legacy of Frederick Jackson Turner', *Western Historical Quarterly*, 18.

Davidson, B. R. (1965) *The Northern Myth: A Study of the Physical and Economic Limits to Agricultural and Pastoral Development in Tropical Australia*. Melbourne: Melbourne University Press.

Day, D. (2005) *Claiming a Continent: A New History of Australia*. Sydney: Harper Perennial.

Dryzek, J. S. (2005) *The Politics of the Earth – Environmental Discourses*. Oxford: Oxford University Press.

Fairclough, N. (1992) *Discourse and Social Change*. Cambridge: Polity Press.

Farwell, G. (1951) *The Outside Track*. Melbourne: Melbourne University Press.

Gammage, B. (2011) *The Biggest Estate on Earth: How Aborigines made Australia*. Sydney: Allen & Unwin.

Head, L. (2000) *Second Nature: The History and Implications of Australia As Aboriginal Landscape*. Melbourne: Angus & Robertson.

Henry, A. D. (2009) 'The Challenge of Learning for Sustainability: A Prolegomenon to Theory', *Human Ecology Review*, 16.

Henzell, T. (2007) *Australian Agriculture: Its History and Challenges*. Collingwood: CSIRO Publishing.

Hill, E. (1937) *Water into Gold*. Sydney: Walkabout.

Hill, R., Golson, K., Lowe, P., Mann, M., Hayes, S. and Blackwood, J. (2005) 'Kimberley Appropriate Economies Roundtable Forum Proceedings'. Cairns: Australian Conservation Foundation, Kimberley Land Council, Environs Kimberley.

Hill, R., Miriuwung and Gajerrong Peoples, Hill, D. G. and Goodson, S. (2008) *Miriuwung-Gajerrong Cultural Planning Framework*. Perth, Kununurra and Cairns: WA DEC, Yawoorroong Miriuwung Gajerrong Yirrgeb Noong Dawang Aboriginal Corporation and CSIRO.

Hill, R., Pert, P. L., Davies, J., Robinson, C. J., Walsh, F. and Falco-Mammone, F. (2013) *Indigenous Land Management in Australia: Extent, Scope, Diversity, Barriers and Success Factors*. Cairns: CSIRO Ecosystems Science.

Hillock, I. M. (2000) *Broken Dreams and Broken Promise: The Cane Conspiracy – Plantation Agriculture in the Northern Territory 1878–1889*. Darwin: NTU Press.

Holmes, J. (2010) 'The Multifunctional Transition in Australia's Tropical Savannas: Emergence of Consumption, Protection and Indigenous Values', *Geographical Research*, 48, pp. 265–280.

Keen, M., Brown, V. A. and Dyball, R. (eds.) (2005) *Social Learning in Environmental Management – Towards a Sustainable Future*. London: Earthscan.

Kinnane, S. (2002) 'Recurring Visions of Australindia', in Gaynor, A., Trinca, M. and Haebich, A. (eds.), *Country: Visions of Land*

and People in Western Australia. Perth: Western Australian Museum.

Kinnane, S. (2008) 'Blood History', in Perkins, R. and Langton, M. (eds.), *First Australians*. Melbourne: The Megunyah Press.

Liberal Party (2013) *2030 Vision Developing Northern Australia* [Online]. Available: www.liberal.org.au/2030-vision-developing-northern-australia

Pascoe, Bruce (2014) *Dark Emu*. Broome: Magabala Books.

Pedersen, H. and Woorunmurra, B. (1995) *Jandamarra and the Bunuba Resistance*. Broome: Magabala books.

Powell, J. M. (1977) *Mirrors of the New World – Images and Image-makers in the Settlement Process*. Folkestone: Dawson & Archon Books.

Powell, J. M. (1988) *An Historical Geography of Australia – The Restive Fringe*. Cambridge: Cambridge University Press.

Pretty, J. (2002) *Agri-Culture: Reconnecting People, Land and Nature*. London: Earthscan.

Pretty, J. (2007) *The Earth Only Endures*. London: Earthscan.

Price, A. G. (1933) 'Pioneer Reactions to a Poor Tropical Environment: A Journey through Central and North Australia in 1932', *Geographical Review*, 23, pp. 353–371.

Robin, L. (2007) *How a Continent Creates a Nation*. Sydney: University of New South Wales Press.

Rose, D. B. (2002) 'Dialogue with Place: Toward an Ecological Body', *Journal of Narrative Theory*, 32, p. 14.

Rose, D. B. (2004) *Reports from a Wild Country: Ethics for Decolonisation*. Sydney: UNSW Press.

Russell-Smith, J., Whitehead, P. J. and Cooke, P. (2009) *Culture, Ecology and Economy of Fire Management in North Australian Savannas: Rekindling the Wurrk Tradition*. Collingwood: CSIRO Publishing.

Sluyter, A. (2002) *Colonialism and Landscape: Postcolonial Theory and Applications*. Lanham, Rowman & Littlefield Publishers Inc.

Sörlin, S. (2002) 'Rituals and Resources of Natural History' in Bravo, M. and Sörlin, S. (eds.), *Narrating the Arctic*. Sagamore Beach, MA: Science History Publications.

Stafford-Smith, M. and Cribb, J. (2009) *Dry Times: Blueprint for a Red Land*. Collingwood: CSIRO.

Swartling, Å. G., Lundholm, C., Plummer, R. and Armitage, D. (2010) 'Workshop Proceedings – Social Learning and Sustainability: Exploring Critical Issues in Relation to Environmental Change and Governance'. Stockholm, Sweden: Stockholm Resilience Centre.

Taylor, G. (1955) *Australia: A Study of Warm Environments and their Effect on British Settlement*. London: Methuen & Co. Ltd.

Wynn, G. (2007) *Canada and Arctic North America – An Environmental History*. Santa Barbara, CA: ABC-Clio.

Yibarbuk, D., Whitehead, P. J., Russell-Smith, J., Jackson, D., Godjuwa, C., Fisher, A.et al. (2001) 'Fire Ecology and Aboriginal Land Management in Central Arnhem Land, Northern Australia: A Tradition of Ecosystem Management', *Journal of Biogeography*, 28, pp. 325–343.

11

Epidemiological regeneration in a complex world

David Waltner-Toews

Chapter 4 outlined a rationale and a theoretical basis for post-normal science in the context of accommodating diverse perspectives in the pursuit of global visions of sustainability and convivial living. As indicated there, these insights emerged from international collaborations in multi- and transdisciplinary research and development projects in Canada, Peru, Kenya and Nepal.

In this chapter, I shall focus on a decade-long series of projects in Nepal that most directly and fundamentally changed the way I thought about, and practised, post-normal science.

The motivations that have brought foreigners to Nepal for more than a century can be grouped into two broad classes: an obsession with climbing big mountains and a search for enlightenment. My motivations have been more complicated. Being a prairie boy afraid of heights, mountains do not interest me and I may, or may not, have found enlightenment.

There is more than one way to tell this story. Indeed, I have spent several decades obsessing over finding the 'right' story. Let me begin this way.

When I first arrived in Kathmandu in 1967 as a callow teenager, Nepal was a quaint, isolated kingdom with a boy king and an alley called Freak Street. Having backpacked and trucked across Europe and Asia, I arrived in Kathmandu as a vagabond, disoriented and culture shocked. I had no particular desire to climb Everest – in fact, I gave away my only good pair of walking shoes to someone who really wanted to go to base camp, but who didn't own decent footwear. Also, having grown up in a religious community, I harboured (and still harbour) a deep suspicion of those who promise enlightenment and was only vaguely interested in what passed for Eastern wisdom.

When I returned to Nepal in the early 1990s, it was indeed part of a lifelong process to determine what 'right action' might look like for a veterinary epidemiologist. In that last decade of the 20th century, Nepal was struggling to become a new democracy, trying to come to terms with an influx of branded clothing, urban sprawl and economic and political chaos. I was there as part of a team of scientists to study and, we hoped, eliminate a parasitic infestation (cystic echinococcosis), which has emerged throughout the world wherever people, livestock (especially sheep) and dogs live in close proximity.

The adult parasites are tapeworms and inhabit the intestines of dogs and other canids, where they mate and send eggs off into the world. If you see them in a pile of dog excrement, they look like miniature, wriggling grains of rice. If people and other animals inadvertently ingest these wriggling tapeworm segments (they are very tiny!), this parasite grows into a large, tumour-like cyst full of baby tapeworms. Surgical removal – one of the few effective treatments, is risky. In Nepal, my colleagues cited a death rate of about 20 per cent for people who went for surgery. This parasite and its relatives are well described in the scientific literature, some of it authored or co-authored by me and it is not my intent to pursue that narrative here. What I will pursue in this chapter is the process and perspective that emerged from, and informed, the work we did.

It was a chilly early morning in 1992 when we walked down the hill and across the footbridge over the Bishnumati River. Fires flickered on the far bank, men and dark shapes of water buffaloes appeared and disappeared into the shadows.

A buffalo, killed by a spike into the back of the skull, sank to the ground as one of the men moved in to slit its throat. In rubber boots, or barefoot, they piled straw over the body and fanned it into flames to burn off the coarse, black hair, then stood back to watch. When the fire died down, they rolled the stiffening corpse onto its back and one of them leaned down to slit open the belly. All along the riverbank, teams of young men waded through spilled guts, stomach contents and faeces, steam rising into the air around them as they quietly went about the business of preparing food for tourists and a rapidly expanding middle class. Dogs and pigs nosed in and out of the fading darkness at the periphery of this tableau. Vultures like great black fruits hunched in the nearby trees.

We were scientists. Not only that, but we were scientists who cared about the practical implications of our work. We knew what to do. Working with Dr D. D. Joshi, a local resident and Director of National Zoonoses & Food Hygiene Research Centre (NZFHRC), we initiated a series of epidemiological studies of the slaughtering practices and the health status and behavioural habits of stray dogs, households and community members. The study was in two wards of Kathmandu, wards being the smallest political administrative unit. Pilot studies by Dr Joshi had shown that many of the animals being slaughtered along the banks of the Bishnumati River adjacent to Wards 19 and 20 in Kathmandu were infected with hydatid cysts. Furthermore, the open-air slaughtering, with dogs and pigs foraging freely, children playing and the use of the river for waste disposal, drinking water, general laundry and bathroom facilities, suggested that this was a high-risk situation for disease transmission.

For three years we did intensive, epidemiological and laboratory research on people, dogs and buffaloes. We devised and delivered public health messages on radio, television and through workshops in the community with various stakeholders.

At the end of these intensive research studies and communications programs, nothing had changed. I was in despair. I had not trained as a veterinarian and epidemiologist to create exquisite models and add exemplary scholarly papers to my curriculum vitae. I had hope of leaving the world a better place than when I'd arrived. And yet, if a team of clever epidemiologists and parasitologists and veterinarians could not even manage to rid one city of one tiny parasite, what was the point of it all?

Flying back to the well-ordered haven of my home in Canada, I dug deeply into my academic training for an appropriate response to what seemed like a world tumbling out of control. I also wrote poetry to fashion a ladder that I hoped would enable me to escape the hole I seemed to be digging ever deeper. I shall return to this in a later chapter.

Around me, diseases and catastrophes were emerging, it seemed, out of nowhere (Lederberg et al., 1992). The old scientific, technical and political agendas seemed powerless – and lacked the economic and moral and intellectual resources – to respond effectively. What were policy makers to do? Set a list of priorities and dollar for dollar go down the list? The list was endless, and even now, after decades of disease-fighting successes, it still is. Furthermore, the ways in which we solved problems of hunger and clean water and unemployment have created conditions for the emergence and spread of new diseases, extinction of species or riots in the streets.

If simple priority setting doesn't work, what are we left with? In conventional development and scientific terms, not much. Researchers can do independent disciplinary work, based on a view of science where every discipline has its own immutable rules of practice, acceptable evidence and quality control. The scholars then hand the results over to politicians, whom we expect to be much more thoughtful, insightful, integrative and visionary than any university scholar. We then lament that the politicians make decisions based on someone else's evidence rather than our own.

Policy makers, even the most thoughtful among them, are facing the prospect of rabies competing with child poverty and maternal health and population control and disappearing wetlands and whales and birds and pesticide pollution and avian influenza and cholera and hunger and childhood obesity and car accidents.

On the verge of a conversion experience I didn't see coming, I decided to go back to basics. Scientists are hunters and gatherers. We gather facts, like fruits and nuts, in a confusing, deep forest we call the world. We do not actually know the world. We perceive with our senses – eyes, ears, noses, tongues, hands – and our clever technical extensions of them. These inputs create a set of neural connections that function as an internal mental model of 'what is out there'. This mental model determines our behaviour – and our survival.

Sometimes individuals are wrong. Sometimes a person's neural connections and biochemical messengers get mixed up, or they take a singular case to be of general applicability, or a person ends up in the wrong place at the wrong time because the landscape on which their brain connections were based has changed. Mistakes can be costly. An animal could be eaten, or run over by a car, or fly into a wind turbine, or a person could eat something toxic or the job for which they have been training for ten years might suddenly disappear. Or an organization or way of doing things can suddenly become irrelevant.

We overcome these deficits by sharing our data and our narratives, by critiquing one another and by adapting to new insights as quickly as we can. We base these stories on personal experience, or a particular study, or a series of studies. This is no more, or less, than a hunting tale recounted around a campfire, a way of constructing meaning from facts. Good science is not just fact finding. It is a systematic, open way to share experience, to offer alternative explanations, to project future possibilities based on past experiences. It links independent observations – chosen by observers according to what they deemed most important or valuable – to generate new insights; insights and actions that transcend the individuals. This is what differentiates a scientist from a pathologically compulsive data hoarder.

This all sounds fine. But how does one do this in practice? This world is a tangle of feedback loops, tipping points and nonlinear dynamics. But the task of understanding our place here is not hopeless. Just as fever and diarrhoea are not separate problems to be solved, but expressions of a body in distress, so our many problems are symptoms of interactions among social and ecological changes.

At this point in my reinvention as a scientist, and working closely with colleagues in Canada, Europe, Asia, Latin America and Africa, I decided that, almost certainly – for there is no certainty – we can have no single true description of our situation. What we have are the multiple perspectives of microbiologists and anthropologists, shamans and epidemiologists, ecologists and economists, farmers, sex workers, theologians and musicians competing for our attention, each group assuring us that they have 'the truth'. For an epidemiologist, this was a problematic discovery. If my research and the quantitative (or at least categorical) models that emerged from it did not offer definitive descriptions of, and solutions to, the diseases I was studying, what was the point?

My colleagues in the journey assured me that, as good scientists and good development workers and good policy makers, we listen and engage with all those people speaking for their different perspectives. The truth, perhaps, would emerge. It all seemed a bit tenuous to me. As a veterinarian, I could see how this post-normal understanding of science might be viewed as a 'scaling up' of what medical and veterinary professionals did every day. The public participants are sources of evidence in the same way that a personal history, a physical examination and a laboratory test inform a clinical medical decision. Post-normal science was not about overthrowing or replacing the many paradigms of science, but of jostling them in such a way that multiple perspectives could be accommodated to generate new insights, generating better diagnostics and responses.

In a veterinary diagnostic setting, observations are integrated to achieve outcomes based on the values espoused by the veterinarian and the animal's owner, such as elimination of a disease in a dog, increased growth and production in farm animals or improved mobility in race horses. When applying these ideas in a community the desired outcomes are not pre-determined. They emerge from a complex process of respectfully accommodating multiple, sometimes conflicting, value-based observations.

Scientific studies over the past few centuries have often assumed that the world was stable, as in a laboratory, that therefore the future could be predicted, and that scientifically based policies, if imposed from above, would have beneficial effects. This is the model most often used in development work and which was the basis for my epidemiological training. It works pretty well in the artificially stable setting of a laboratory, or even within the bounds of hospitals or industries. In the complex world we inhabit, not so much.

This is what we had tried in Kathmandu and which failed. We could create a scientific model to predict that if you got rid of the street dogs, slaughtered all the buffaloes in a government-inspected abattoir and had a professional cadre of sanitary engineers, Nepal would be seen, in this narrative, to be a healthier place. But shopkeepers and priests didn't want to get rid of the street dogs because they were acting as community police. Butchers didn't want to work in a state-of-the-art slaughterhouse as paid employees of a wealthy owner because butchering was a family tradition. The women who cleaned the streets wanted day care for their babies and education for themselves so that they could properly engage in public debates over waste management. And democratic change can be chaotic and animal inspection laws are not the highest priority of new governments.

Nepal is not unique. The experts can tell us not to smoke, to eat better, to exercise, to not let our dogs or chickens run in the street, to cycle and recycle, yet in the full knowledge of what we are doing, we resist changing our behaviour. This is not always the result of ignorance remediable through more finely constructed communication messages. We all have defensible reasons for doing what we do, based on the particular way we simplify that opaque and complex world out there, the models and lenses we use to understand it.

Many of us have some idea what a scientific experiment might look like. But what might a hybrid experiment look like? A model where no one is an authority and the expertise is distributed among many people, where policies are the hypotheses and programs are the research, and where the outcomes need to be monitored from various perspectives so that actions can be adapted or altered?

In the mid-1990s, I returned to Nepal. The two partners involved in the first study, together with Canada's International Development Research Centre (IDRC), which was funding the work, recognized the need for local expertise in social action and mobilization. Consequently, a local nongovernment organization, Social Action for Grassroots Unity and Networking (SAGUN) was invited to be a part of the research team, to complement the epidemiological and biomedical skills of the collaborators from the University of Guelph and the NZFHRC. Our new research team included community activists, butchers, squatters, shopkeepers, politicians, anthropologists, epidemiologists and school children. Each of us brought our own values and perspectives. At the time my own values were somewhat inchoate, but, in retrospect, they would look a lot like the feminist politics of care recently articulated by Wickson et al. (2017).

The specific objectives that emerged from an October 1997 Planning Workshop in Kathmandu, hosted by the two local nongovernment organizations (NGOs) and IDRC, were the following:

- to increase the technical understanding of Kathmandu's urban ecosystem (on the part of both researchers and community members)
- to enhance the capacity of individuals and organizations to improve currently poor public health conditions, articulate demands, and lobby for change through the combination of the ecosystem approach and participatory action research.

Managing cystic hydatidosis, the highest priority for the academic researchers, was not explicitly mentioned. Within three years, the community had transformed its neighbourhoods, creating parks and gardens and enclosed latrines along the riverbank, building small-scale abattoirs, composting the stomach contents from killed animals, raising ducks on the grubs growing in the compost and selling the final product as fertilizer. They were thinking and working systemically to create a complex, multilayered narrative of which, so they explicitly declared, their grandchildren could be proud (Waltner-Toews et al., 2005). In the process, as a side effect, the conditions that had facilitated the maintenance of the parasite disappeared.

Those of us from a health science background could frame the emergent outcome using the language with which we were familiar: that of health. The World Health Organization has defined health as a state 'of complete physical, mental and social well-being'. What we saw emerging from this work, however, was closer to what microbiologist Rene Dubos defined as 'a modus vivendi enabling imperfect men to achieve a rewarding and not too painful existence while they cope with an imperfect world' (Rene Dubos, as cited in Last, 1988). Or the definition suggested by Ivan Illich, who wrote that health 'designates a process of adaptation. It designates the ability to adapt to changing environments, to growing up and aging, to healing when damaged, to suffering and to the peaceful expectation of death. Health embraces the future as well, and therefore includes anguish and the inner resources to live with it ... Health is a task' (Illich, 1975, pp. 105–106).

Having said that, I also became aware that this emergent outcome could be described in other terms, such as resilience or sustainability. Just as each of us brought our own value to the table, we also framed what we did in the languages with which we were familiar. The outcome was, and is, more important than the language used to describe it.

In parallel with our work in Nepal, we collaborated on projects in South and Southeast Asia, Latin America, East Africa and Canada. What emerged for us as researchers was not a set of rules for doing scientific experiments or epidemiological studies in the field, but a set of guidelines. These guidelines were given various names; we called ours an Adaptive Methodology for Ecosystem Sustainability and Health or AMESH for short. As with similar initiatives emerging from other theories and practices, AMESH offered a kind of 'rough guide' and proposed a circular series of seven steps and guiding questions (Waltner-Toews and Kay, 2005):

1. What is the presenting complaint that has brought us to this place?
2. Who are the participants in this situation? What do they care about? What rules govern their behaviour and decisions? Where do non-human species fit into this?
3. What are the stories told by the participants about how the situation came to be and their role(s) in it?
4. What are the scientific, social and cultural issues that need to be addressed and how are they related to each other?
5. What are the scenarios, visions or narratives that are both feasible and desirable?
6. Can we design a programme to achieve our common goals? Who will implement this programme?
7. What kind of a monitoring system will enable us to measure progress? Who will do this? How will that information be fed back into the system, and how can we use that information to adjust our programmes?

Each of these steps was associated with more questions that probed deeply into relevant issues (Neudoerffer et al., 2008). There was no recipe, but there is the excitement of working with other people to find more convivial ways to live on the planet.

This ecosystem approach, whether applied to health or conservation or sustainable development brings together our best understanding of the historical, cultural and ecological complexity of the world with an emerging goal that transcends culture and history. We have found that this process and its variants open a space to struggle with tragedy, suffering and contradiction and for integrating interventions and policies for health, biodiversity, agriculture, social equity and local democracy.

As with all local initiatives, the global context mattered, creating the necessary, although insufficient, conditions for local action. The fall of the Berlin Wall in 1989 opened the door to new, activist policies in the Butchers' Association in Kathmandu. These policies, as they were implemented, made a space for the kind of transformational change we saw on the banks of the Bishnumati. It was not incidental to our success that, set in a place of honour at the Butchers' Association meeting hall, were pictures of two young martyrs to the democracy movement. Neither was it incidental that the working policy of my primary collaborator, Dr D. D. Joshi, was based on a commitment to improving the neighbourhoods in which he lived, no matter who held political power.

In 2011, American policies related to the War on Terror coincided with multiple murders of members of the royal family and led to increased turmoil and open civil war in Nepal. Yet the Kathmandu neighbourhoods with whom we worked not only survived, but their ideas have been extended to other communities. Their policies were positioned to be adaptive and take advantage of, rather than be trapped by, changing social–ecological conditions. What the people with whom I worked in Kathmandu taught me is that, if we draw on a wide net of human experience, policies can anticipate several possible futures and be prepared for them.

In the process of working with my friends and colleagues in Nepal, I realized that the work we were doing together, imperfectly, was more than just a new way to think about science, knowledge creation, health and problem solving. It can serve to speak across cultures. One of the biggest challenges is to find a language for this. Descartes and his contemporaries hoped that science could offer a sort of universal language, a kind of Esperanto based on the Latin of upper-class European elites. But normal scientists, after enabling us to achieve remarkable technological feats and to overcome the threats of deadly plagues that once decimated human and animal populations, have brought us finally into an overwhelming, obsessively detailed swamp of 'big data'. We can name the species, but not the despair or find words for a way forward. Joshua Lederberg declared that biology 'is already so fact laden that it is in danger of being bogged down awaiting advances in logic and linguistics to ease the integration of the particulars' (Lederberg, 1995).

At a 2010 workshop on an emerging epidemic of canine rabies in Bali, Indonesia, we struggled to find ways of working together that were more than (other than) what might seem to be just another transplanted European idea. As we struggled to find the words, Drh Anak Agung Gde Putra, a remarkable scientist, administrator and traditional prince, talked about Hinduism in Bali, and whether we might find something there. Tri Hita Karana, he informed us, comes from one of the Hindu holy books and advises seeking happiness in the here and hereafter through a harmonious relationship with God, with fellow humans and with nature. Tri Hita Karana is but one version of what many cultures throughout the world have known and in the past century has been expressed by academic researchers as transdisciplinarity and by those working at the interface between science and policy as post-normal science. Transdisciplinarity, a vision which transcends our imperfect disciplinary understanding, seems to me to be a kind of secular version of a harmonious relationship with God. Participation, incorporating economic, gender and ethnic equity, was an expression of harmonious relationship among humans. Sustainability and its intellectual relatives represent a harmonious relationship with nature. What we were doing, and calling ecohealth, in many ways, could be seen a scientific and political reformulation of an ancient and deep human insight.

In October of 2010, I participated in a workshop puzzling over the relationship between biodiversity and indigenous health in the Amazon and the eastern slopes of the Andes, the Yungas. The workshop included physicians, anthropologists, epidemiologists, veterinarians, teachers, ecologists, naturalists. We were indigenous peoples, Argentinians, Canadians, Brazilians, Brits, Peruvians, Colombians. Over a week, we argued, drew pictures, pored over maps, sang, danced, yelled at one another, cried, laughed, got drunk, checked with our social networks on Skype, made lists and organizational diagrams, came back to the table and drew up a research plan. What we were forging was an uneasy sense of common vision, a vision of imperfect people trying to achieve 'a rewarding and not too painful existence while they cope with an imperfect world'. At several points the debates and arguments were heated. One indigenous leader suggested that their traditional way of dealing with those with whom they radically disagreed was to kill them. This workshop was outside their comfort zone. Well, this was outside the comfort zone of many of us. And yet, I thought, if she wanted to kill us, that meant that we were dealing with subject matter that was deeply important to her and her people. The fact that she didn't kill us gave me hope that a convivial future was possible and that the post normal, open-ended, mutually respectful journey we had embarked offered the possibility of a way forward.

Acknowledgements

The author acknowledges the support of Canada's International Development Research Centre for much of the research described in this chapter. An earlier version of this chapter was presented on 24 November 2010, as part of a 'Speakers of Renown' series at IDRC's headquarters in Ottawa. Celebrating IDRC's 40th anniversary, this series brought together 'the best minds in international development for free public lectures on a wide range of pressing issues'.

References

Illich, I. (1975) *Medical Nemesis.* London: Calder & Boyars, Ltd.

Kuhn, T. (1962) *The Structure of Scientific Revolutions.* Chicago: University of Chicago Press.

Last, J. (1988) *A Dictionary of Epidemiology.* Oxford: Oxford University Press.

Lederberg J. (1995) *Preface to Haldane's Daedalus Revisited.* Dronamraju, K. R. (ed.). Oxford: Oxford University Press.

Lederberg, J., Shope, R. E. and Oaks, S. C. (eds.) (1992) *Emerging Infections: Microbial Threats to Health in the United States. Institute of Medicine.* Washington, DC: National Academy Press.

Neudoerffer, R. C., Waltner Toews, D. and Kay, J. J. (2008) 'Return to Kathmandu: a Post-Hoc Application of AMESH', in Waltner-Toews, D., Kay, J. and Lister, N.-M. (eds.), *The Ecosystem Approach: Complexity, Uncertainty, and Managing for Sustainability.* New York: Columbia University Press.

Waltner-Toews, D. and Kay, J. (2005) 'The Evolution of an Ecosystem Approach: the Diamond Schematic and an Adaptive Methodology for Ecosystem Sustainability and Health', *Ecology and Society*, 10(1), p. 38;http://www.ecologyandsociety.org/vol10/ iss1/art38/.

Waltner-Toews, D., Neudeorffer, C., Joshi, D. D. and Tamang, M. S. (2005) 'Agro-urban Ecosystem Health Assessment in Kathmandu, Nepal: Epidemiology, Systems, Narratives', *Ecohealth*, 2, pp. 1–11.

Wickson, F., Preston, C., Rosa Binimelis, R., Herrero, A., Hartley, S., Wynberg, R.et al. (2017) 'Addressing Socio-Economic and Ethical Considerations in Biotechnology Governance: The Potential of a New Politics of Care', *Food Ethics*, 1, pp. 193–199.

12

Landscape management and landscape regeneration in Australia

Richard M. Thackway and Mark W. Gardner

Introduction

For almost 160 years almost 60 per cent of Australia's landscapes have been managed primarily to produce food and fibre (see Table 12.1). Covering almost 7,687,000 km^2, there is widespread evidence almost everywhere to show the ecological effects that rural production and pest animals and plants associated with agricultural industries have had on the environment (Saunders et al., 1990). At the same time, farming has been a major contributor to the Australian economy. In the first half of the 20th century, agricultural production accounted for up to 80 per cent of Australia's exports and around a quarter of the nation's output (Productivity Commission, 2005). Development of the environment for agricultural production has come at a cost to the environment. Large areas of Australia have been, and continue to be, modified and degraded. This adversely affects the resilience of these once healthy functioning ecosystems (Metcalfe and Bui, 2017; Thackway and Freudenberger, 2016). This is regrettable, since sustainable agricultural production depends on the inherent healthy functioning of these ecosystems. Few areas of landscapes in Australia are managed with the aim of enhancing, maintaining and protecting the regenerative capacity of healthy functioning ecosystems.

Table 12.1 Proportion of Australia managed for agricultural enterprises, 2005–06

Land use	*Area (km²)*	*Proportion of Australia managed for agricultural enterprises (%)*
Grazing native vegetation	3,558,785	46.30
Grazing modified pastures	720,182	9.37
Dryland cropping	255,524	3.32
Irrigated cropping	12,863	0.17
Irrigated pastures	10,011	0.13
Irrigated horticulture	3,954	0.05
Intensive animal and plant production	3,329	0.04
Dryland horticulture	1,092	0.01
Total	**7,687,147**	**59.39**

Source: ABARES, 2016

Industrial agriculture and the Anthropocene are part and parcel of the same phenomenon. The worldwide modification of ecological function can be attributed to a combination of either inadvertent or deliberate land management regimes that affect the function, structure and composition of the agricultural landscape. These modifications are obvious to the trained eye and include all landscapes. There are the intensively managed higher rainfall zones in the southern and eastern margins of the Australian continent. Ecological modifications can also be observed in the extensively managed lower rainfall zones found in the semi-arid and arid hinterland.

Industrial agriculture is characterized by, and limited to, the broad-scale use of the following management regimes: converting complex natural ecosystems to monoculture crops; diverse pastures and plantations; heavy reliance on fossil fuels in combustion engines and inorganic fertilizers; widespread use of ploughing and fallowing that emit rather than store carbon. We have also observed the widespread acceptance that overgrazing of Australian rangelands and their exposure to climatic extremes is a normal part of the means of production in these semi-arid ecosystems. There is also an unwillingness to change from the use of inefficient and deleterious irrigation systems, which is not only a profligate waste of water, but also has caused considerable losses of productivity and biodiversity through raising saline water tables (Metcalfe and Bui, 2017).

Compared to industrial agriculture, regenerative agriculture aims to establish land management regimes that capture and put more long-lasting carbon into the soil, while minimizing the loss of soil carbon. This approach can reverse the harmful carbon emitting effects of conventional industrial agriculture and positively influence the local water cycle, which is crucial for moderating local, regional and national climate patterns (Massy, 2017).

Regenerative agriculture is an emerging discipline. Our observations across Australia among practitioners of regenerative agriculture show there is a diversity of ecological, social and economic drivers related to their means of agricultural production. The types of production system involved include extensive grazing production systems in most agro-climatic regions and intensive pasture and cropping systems mainly in the southern temperate regions, as well as in coastal environments in the tropics and subtropical regions.

The aim of this chapter is to present some emergent observations from applying standardized survey instruments at the farm and landscape scales, working in partnership with regenerative landholders. We discuss these observations of regenerative landscape managers and their transformed social, financial and ecological outlooks. We document and discuss the opportunities for these regenerative initiatives to represent viable sustainable agricultural enterprises. We also consider the potential for upscaling these findings to the broader agricultural sector and comment on some of the likely public–private benefits of implementing regenerative landscape management regimes Australia wide.

Associated with regenerative agriculture are myriad innovative regenerative land management practices. In cropping landscapes, these innovations include: full-cover cropping; multi-species cover cropping, no-kill cropping, pasture cropping and biological cropping. There is also a range of combinations of each, as well as no-till operations (Massy, 2017; Norton and Reid, 2013). In grazing landscapes, these innovations include: using combinations of cattle and sheep to incorporate organic matter into the soil through trampling; using perennial pastures to capture rainfall where it falls; slowing overland flow of surface runoff associated with intense rainfall events; repairing eroded landscapes by holding water higher in the landscape; re-establishing leaky weirs and chains of ponds in eroded creek lines; and protecting the soil surface from rain splash erosion; (Massy, 2017; Norton and Reid, 2013).

Depending on the land types, agro-climatic regions and enterprises, we have also observed many regenerative practices that are becoming increasingly recognized as means for restoring, repairing and improving the ecological function of the land. These include:

- plant-based approaches such as revegetation and rehabilitation of degraded areas and restoration of vegetative structure and composition in denuded and eroded areas through low cost methods focusing on using animals and planned grazing

- soil-based approaches such as applying organic fertilizers and ameliorants to re-establish healthy living soils; re-establishing appropriate ratios of bacteria and fungi; encouraging soil bioturbators; capturing and using rainfall where it falls through ground cover management rather than dispelling and concentrating rainfed surface flows into gullies, channels and streams
- biodiversity-based approaches such as encouraging and protecting habitats for birds, insects and invertebrates to assist in insect pest control and in creating conditions for plant succession.

These and other nature-based systems of land management practices that are used, particularly in the developing countries, are described by Jordan (1998).

The number of regenerative landscape managers involved in the production of food and fibre is relatively small, compared to Australia's estimated 135,000 farmers and graziers (ABS, 2012). Arguably, the number of these farmers could be greatly increased with the development of land-use and management policies together with incentives for the widespread adoption of regenerative landscape management regimes.

Defining and assessing whether landscape management is regenerative

The main determinate for a farmer/grazier to be identified as regenerative is how they think about the land and their relationship with it (Massy, 2017). We also acknowledge the importance of a holistic approach to the whole-of-farm includes the whole family and their aspirations (including financial), as well as the biophysical environment they manage, including socioeconomic, ethical, aesthetic and sympathetic attributes of living on the land as a farmer. While these dimensions are vital and an important part of regenerative management, our focus here is assessing whether a landscape is regeneratively managed and then to assess the flow-on effects on production, financial and social outcomes.

The ecological status and performance of a natural ecosystem's condition can be monitored, assessed and reported using ecological criteria and indicators of function, structure and composition (e.g. MIGNFISC, 2013; Noss, 1990; Thackway and Freudenberger, 2016; Thackway and Specht, 2015). Similarly, the ecological status and performance of agricultural landscapes on farms can be assessed using ecological response criteria and indicators that include function, structure and composition. An assessment of these responses, in farming and grazing enterprises, can determine the extent to which a land manager and their property is an expression of regeneratively managing the landscape.

The concept of regenerative landscape management is increasingly being written and spoken about (Marsh, 2018; Massy, 2017; White, 2012). However, there is little in the scientific literature to define this important management philosophy.

We have developed a working definition as follows:

> Regenerative farming and grazing systems are holistically managed landscapes that show a consistent positive response, or a consistent high functioning stable pattern across the ten criteria (see Table 12.2) over time to landscape management practices and the harvesting of food and/or fibre. By nature, they should require a low level of inputs to do this.

This approach, which includes the definition and the associated conceptual model and assessment framework (see Table 12.2 and Figure 12.1), has been applied across a mosaic of agricultural ecosystems; including landscapes

Table 12.2 Hierarchy of five response variables of the environment and biodiversity (i–v) crossmatched to ten ecological response criteria (1–10). This assessment framework has been developed to operate in intensive and extensive agriculture and minimum use areas, at the land type level

Five response variables of the environment and biodiversity	*Primary ecological response criteria*	*Secondary ecological response criteria*
i Manage land types to minimize effects of extreme climatic events, which considers the whole property and its place in the broader catchment	1 Resilience to natural major climate events (e.g. drought, wildfire, flood, dust storms, cyclone)	
ii Manage soils to prevent erosion and to maintain ecological health, productive capacity and water quality of the property and watershed	2 Status of soil nutrients including soil carbon, major and minor elements	
	3 Status of soil hydrology including infiltration, percolation and water availability to plants	
	4 Status of soil biology including bioturbators, i.e. nutrient recyclers, fungi and bacteria ratios and soil organic matter	
	5 Status of soil physical properties including bulk density and soil as a medium for plant development and growth	

iii Manage pastures for production and to maintain ecological health of the property and watershed	8 Status of ground layer/ground cover/grass and herb structure	6 Status of the reproductive potential of the plant species and plant community
	10 Status of the ground layer/grass and herb species richness and functional traits	6 Status of the reproductive potential of the plant species and plant community
iv Manage trees and shrubs for production and to maintain ecological health of the property and watershed	7 Status of tree and shrub structure	6 Status of the reproductive potential of the plant species and plant community
	9 Status of tree and shrub species richness and functional traits	6 Status of the reproductive potential of the plant species and plant community
v Manage natural watercourses, riparian areas, natural lakes and wetlands, to protect ecosystems that are sensitive to agricultural land management	6 Status of the reproductive potential of the plant species and plant community	2 Status of soil nutrients including soil carbon, major and minor elements
		3 Status of soil hydrology including infiltration, percolation and water availability to plants

Sources: Classes i–v are modified from Table 2 in McIntyre et al. (2000) and classes 1–10 are modified from Thackway (2014)

and farms that are intensively managed (higher inputs/outputs) as well as in landscapes and farms that are extensively managed (lower inputs/outputs) (Thackway and Specht, 2015; Thackway and Freudenberger, 2016; Yapp and Thackway 2015). Some of these properties that we have assessed have no or only a very small area set aside and are specifically managed for the conservation and the protection of biodiversity, with most of the land area being managed for food and/or fibre production using regenerative management regimes. While other properties, in addition to being managed for food and/or fibre production using regenerative management regimes, have a relatively large area set aside and managed specifically for the conservation and protection of biodiversity.

Using this approach, the management practices associated with regenerative farming need some appraisal.

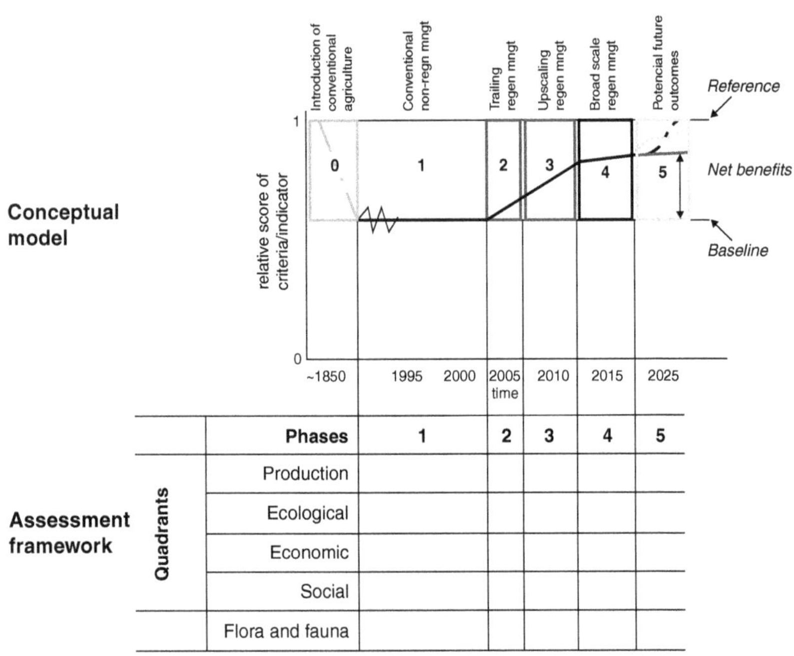

		Phases	1	2	3	4	5
Assessment framework	**Quadrants**	Production					
		Ecological					
		Economic					
		Social					
		Flora and fauna					

Figure 12.1
Conceptual model and assessment framework to guide which outcomes of regenerative management are to be assessed (modified from Thackway 2013; Yapp & Thackway 2015)

In a regenerative approach, reliance on inputs should diminish over time, as the capacity of the ecosystem to self-organize becomes strongly established (Massy, 2017).

Applying a standardized approach

Our research approach is based on the associated conceptual model and assessment framework (Figure 12.1 and Table 12.2) and standardized survey instruments that are consistently and repeatedly employed over time. This approach guides the design of projects and assists in determining what outcomes are to be assessed. The farmers' management ideals and goals for their regenerative landscape management journey are documented as phases over time.

This approach has been applied to the past as well as the future. In addition to the four main quadrants (i.e. production, ecology, financial and social), an additional element, biological assessments (i.e. flora and fauna), may be added, depending on the requirements of the manager. At the farm level, these quadrants are synthesized to reveal emergent properties about a 'farmer's compelling story'; of how the farmer thinks about the land and their relationship with it. This holistic approach also seeks to reveal the whole family's aspirations, the biophysical environment they manage including the socioeconomic, ethical, aesthetic and sympathetic attributes of living on the land as a farmer.

Standardized survey instruments for each of the four quadrants are consistently applied to collect, assess, track and report socioecological and financial changes and trends. Fundamental to understanding and assessing and interpreting changes and trends involves working with the land managers to understand their priorities (goals) and management ideals (values) during the different phases of the adoption and implementation of regenerative land management regimes; that is, phase 1 (before), phases

2 and 3 (during) and phase 4 (after) (Figure 12.1). Two additional phases assist to bookend the adoption and implementation of regenerative land management regimes: phase 0 provides a historical context and phase 5 considers potential future outcomes and shows indicative trajectories that may also be possible through the development of innovative regenerative management regimes.

The standardized approach is underpinned by the ability to evaluate and benchmark production, ecological, economic and social outcomes against independent datasets.

Understanding the goals of regenerative farmers

While the concept of regenerative landscape managers is commonly discussed from an ecological perspective, focusing solely on the biophysical neglects the impact of the approach on the people living and working on the land as well as the effect on farm profit. In addition to an ecological assessment, our research involves assessing production, economic and social aspects of regenerative landscape management in agricultural settings.

It has become obvious to the authors that we have become mutual learners in the gathering of information. It is evident to us that the movement towards regenerative farming is a transformative change, as opposed to simply learning more of the same and doing the same things incrementally more efficiently. This opens the possibility of changes to the thinking patterns (and thought processes) of those involved, including the gatherers of information, the regenerative farmer and their family. Flowing on from this mutual co-learning is the potential for this knowledge to influence changes in the relationships of the researcher, individuals, families, society and their relationships to their environment.

In the example in Table 12.3, data collected by Mark Gardner and Sue Ogilvy (Fenner School of the Environment

Table 12.3 Standardized list of 24 priorities (goals)

1	Maximum income	13	Good crops, pastures, and stock
2	Satisfactory income	14	Farm as a business
3	Keeping out of debt	15	Important to the community
4	Safeguarding income	16	Leaving farm in good/better condition
5	Gaining recognition	17	Improving biodiversity
6	Family tradition	18	Special abilities and aptitudes
7	Being own boss	19	Meeting a challenge
8	Children in worthwhile occupations	20	Enjoyment of work
9	Farm in the family	21	Making farm productive
10	Working with family members	22	Healthy outdoor life
11	Pride of land ownership	23	Working hard
12	Self-respect	24	Independence

Source: Vanguard Business Services On Track Farm Family Social Indicators

and Society and Team Leader for an Australian Government's National Environmental Science Program), a large sample of regenerative farmers were asked a series of standardized social survey questions and asked to select the most important goals from the list of 24 options. They were then asked to prioritize five choices in importance, from 5 to 1. This sample was a subset of farmers known to be regenerative landscape managers.

Emerging themes

Goals and priorities

The results from social survey outlined in Table 12.3 can give us insight into the goals of farmers and graziers. This can assist to better understand priorities and motivations.

In the survey, in the grassy woodlands of NSW the three highest scores for the goal statements (shown in brackets) were:

Goal 16: leaving the farm in good or better condition (67)
Goal 17: improving biodiversity (52)
Goal 2: achieving a satisfactory level of income (52)

Note the highest score possible for any one goal is 120.

It is not surprising that the highest priorities of the regenerative graziers were to 'leave their farm in good or better condition' and to 'improve biodiversity'.

In this group, the most important financial goal was to achieve a 'satisfactory level of income' (52). It is also noteworthy that 'maximizing income' scored only 4. The surveyed information was reinforced in the researchers' facilitated conversations, as the participants spoke about achieving 'enough' profit to sustain their requirements (social, physical and financial) rather than to extract the maximum profit from the resources (and hence potentially conflict with their two most important goals).

These findings show an important distinction between regenerative graziers and conventional approaches to farming, where in the latter approach to farming there is a greater emphasis placed on profit and production maximization (Holmes and Sackett, 2016).

Regenerative graziers have a clear emphasis in their management on improving and maintaining landscape health even though it could be expected that farm profit would suffer as a result. Soon to be published research completed by Mark Gardner and Sue Ogilvy has shown that the profit achieved by the survey respondents is broadly comparable with two benchmark groups: the average producers in the Holmes and Sackett AgInsights and the producers in the MLA mixed farming project. This is of great interest, because such producers are described in the Holmes and Sackett report as being highly profit and production motivated. Regenerative graziers, despite not having the same focus on profit, were able to achieve a comparable level of profit as a by-product of their land management approach (i.e. regenerative land management regimes). Mark Gardner and Sue Ogilvy continue to work in this area.

It is important to note that further investigations and comparisons showed major differences in production, income and cost structures between the two groups.

An emerging ethical, aesthetic and sympathetic relationship with the land

Most producers involved in regenerative landscape management, while having a background in industrial or conventional agriculture, have opted to develop a more holistic relationship with their land, which includes the way they and their family think about the land and their relationships with it (Massy, 2017).

Regenerative land managers are showing that degraded soils can be restored (e.g. Soils for Life, 2016). A high level of soil carbon indicates a resilient and productive agricultural landscape, while farms without such a level of carbon cannot sustainably continue to produce food and fibre. Successful regenerative landscape managers have developed a sound understanding of soil hydrology, nutrients, physical properties and biology. In addition, successful regenerative farmers have also developed an understanding of climate variability and its effects and the vulnerability of their enterprises to a changing climate.

Adaptation and resilience

Producers involved in regenerative farming and grazing note that their production systems have typically less reliance on external inputs and produce lower outputs. They also observe that the agricultural ecosystems that they manage eventually develop new stable states, where the indicators of ecological functions (including soil and vegetation condition and water quality) are encouraged to be self-organizing.

The role of regenerative farming practices may be to confer a greater capacity and certainty of the ecosystem and the farm business to adapt to changing climate patterns and therefore become better suited to future extremes. Agricultural ecosystems and farm businesses that have been regarded as resilient may not necessarily have the capacity to bounce back under such extreme disruptions such as we are seeing from the current drought across much of eastern Australia.

Example of regenerative landscape management

One example of regenerative landscape management is on a grazing property on the Southern Tablelands of NSW. The property manager utilizes a holistic management grazing chart to plan and record flock movements across each paddock on the farm for the growing and non-growing season. In this example, we collated and analysed the land managers' records for the period 1999 to 2016.

Figure 12.2 shows the changes in carrying capacity over time for the paddocks. To dry sheep equivalent rating (DSE), carrying capacity was calculated by determining the number of animals and converting that number to DSE using standardized NSW government conversion charts. The number of days animals were in a paddock throughout the year was considered to give an overall paddock stocking rate of DSE days/ha/year. Figure 12.2 also shows a linear trend line since 1999, over which there has been a steadily increasing number of available grazing days.

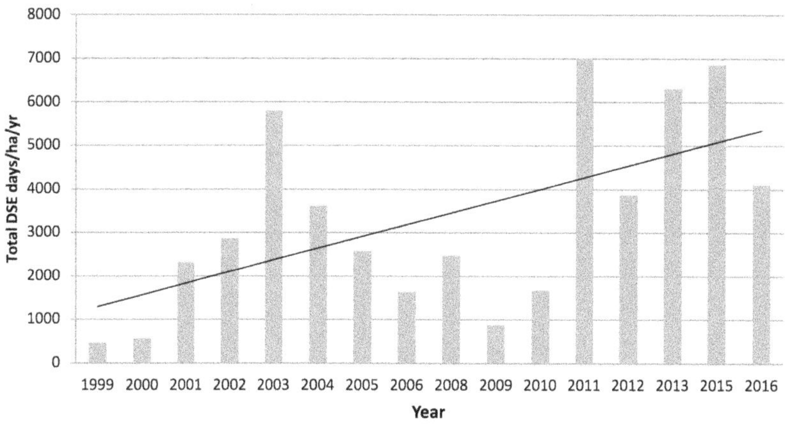

Figure 12.2 **Grazing plan analysis (total DSE days/ha/yr) with a linear trend line added**

A wildfire in summer 2006 burnt significant areas of the property and reduced the available pasture over the following four growing seasons. The impact of the removal of ground cover can be seen for some four years (2006–10). During the recovery phase, paddocks were grazed and then allowed long recovery periods, more than 150 days, reducing DSE days/ha for these periods. The resilience of the paddocks is evident, with total DSE days/ha recovering strongly.

We combined the financial return for this paddock with all paddocks on this property to determine the financial benefit associated with managing the property using regenerative land management practices (see Figure 12.3). This figure shows there is a decline in profit corresponding to the period (2006–10) during which ground cover was very sparse but recovering following the wildfire.

The response of ecological functions over time were assessed using the Vegetation, Assets, States and Transitions (VAST) ecological analysis (Thackway and Freudenberger, 2016; Thackway and Specht, 2015). The land manager works with VAST to diagnose, account for, monitor and report on the interactions between the biophysical effects of land management on transforming landscapes over time (Figure 12.1 and Table 12.1).

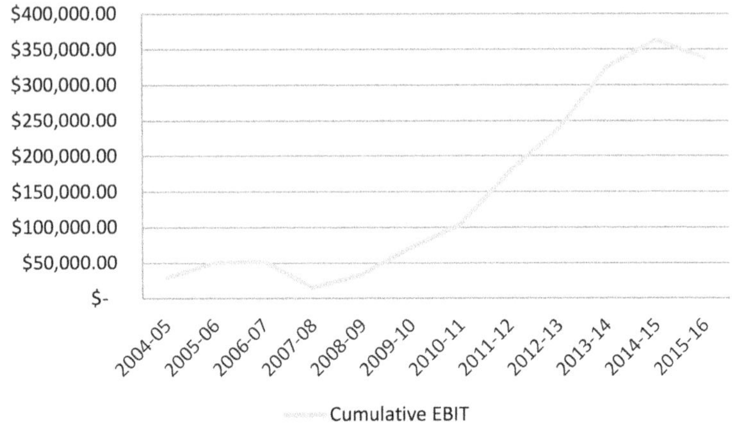

Figure 12.3 **Cumulative earnings before interest and taxes (EBIT) for the whole property**

Based on an ecological assessment for this property over time (1965–2017) several observations were also made by the land manager relative to a 1965 baseline:

- The number of grass species were observed to increase naturally in the regenerating pasture without being planted. Over this period the number of native grasses increased from seven to 58 species. It is believed that the additional species recorded were initially present in the soil and the surrounding fence paddock lines, but in such low populations and cover as not to be initially observed.
- Soil fungal activity was observed to be initially very low because of intensive land management including cropping, weed control, the additional of chemical fertilizers and an intensive set stocking grazing regime. The 2006 wildfire appeared to cause a rapid decrease in fungal activity. By 2008 the fungal activity was observed to increase again. That increased steadily until 2015 when it plateaued. The rapid increase in fungal activity was attributed to stock and pasture management; that is, spelling the paddocks after grazing and allowing the pasture to recover over a relatively long period.

- Tests of soil carbon in the same paddock used to report DSE (Figure 12.2) were compared with a district soil carbon database. While many of soil test results for surrounding properties showed soil organic carbon levels of around 1 per cent (associated with isolated paddock trees), the results for this property revealed carbon levels of 3.75 per cent.

Such paddock-based and farm-based assessments show the importance of ecological learning and literacy, of a willingness to experiment and trial innovations and to upscale and implement regenerative landscape management regimes and practices on a whole-farm scale. This progression takes time and resources. Such examples help to illustrate that under a regenerative farming approach to land management it is possible to manage for ecological, production and profit outcomes simultaneously. As noted by Jordan (1998), rather than abandoning current mainstream agriculture and forestry, land managers should aim to adopt a gradual replacement of energy-intensive, high input–output production systems with land management practices that are based on understanding and working within the limits of nature's ecological services.

Conclusions

By allowing nature's services to self-organize, there is great potential to develop production systems that do not rely on high-cost inputs–outputs that fight against the forces of nature. The financial and social benefits of this could be substantial.

There is an emerging understanding among regenerative land managers of the importance of developing and implementing innovative land management regimes to restore, maintain and mimic the natural cycles found in unmodified ecosystems including water, soil, vegetation and biodiversity.

To ensure that Australia's agricultural enterprises are world leading, it is imperative that we all learn from these innovative regenerative farmers. Working beside these famers creates a co-learning environment where the researcher and the farmers learn together in an equal relationship.

Once ecological impacts are considered, we question the economic viability of the current model of industrial agricultural production that pervades the agricultural Australian landscape.

Currently, a vision for broad-scale regenerative landscape management is countercultural. The current dominant industry paradigm is strongly based on production, involving applying an increasing level of industrial chemicals by high-input farmers, with little ecological consideration. This paradigm is supported by publicly funded research and development programmes and chemical retailers, which invest little into understanding and promoting regenerative farming.

Acknowledgements

The authors acknowledge the use of farm and paddock information that was provided by an anonymous land manager and the findings of a research project supported through funding of the Australian Government's National Environmental Science Program and jointly led by the second author and Sue Ogilvy. We gratefully acknowledge the ANU Fenner Writing Group for its editing and reviewing of an earlier draft of the chapter.

References

ABARES (Australian Bureau of Agricultural and Resource Economics and Sciences) (2016) 'National land use of Australia, 2016'. www.agriculture.gov.au/abares/aclump/land-use/land-use-mapping accessed on 31 July 2018.

ABS (Australian Bureau of Statistics) (2012) 'Australian farming and farmers'. 4102.0, *Australian Social Trends*, December 2012. www.abs.gov.au/AUSSTATS/abs@.nsf/Lookup/ 4102.0Main+-Features10Dec+2012 accessed on 31 July 2018.

Holmes and Sackett (2016) *AgInsights Knowing the Past Shaping the Future*, Vol 18. Wagga Wagga: Holmes Sackett and Associates.

Jordan C. F. (1998) *Working with Nature, Resource Management for Sustainability*. Amsterdam: Harwood Academic Publishers.

Marsh, D. (2018) 'Not yet a tsunami, but a gathering tide', 15July. David Marsh Blog, Alliance for Regenerative Landscapes and Social Health. arlash.com/blog/ accessed on 31 July 2018.

Massy, C. (2017) *Call of the Reed Warbler: A New Agriculture – A New Earth*. St Lucia: University of Queensland Press.

McIntyre S., McIvor J. G. and MacLeod N. D. (2000) 'Principles for sustainable grazing in eucalypt woodlands: Landscape-scale indicators and the search for thresholds', in Hale P., Petrie A., Moloney D. and Sattler P. (eds.), *Management for Sustainable Ecosystems*. Centre for Conservation Biology, Brisbane: University of Queensland, pp. 92–100.

Metcalfe, D. J. and Bui, E. N. (2017) *Australia State of the Environment 2016: Land*. Independent report to the Australian government minister for the Environment and Energy, Canberra: Australian Government Department of the Environment and Energy.

MIGNFISC (Montreal Implementation Group National Forest Inventory Steering Committee) (2013) *Australia's State of the Forests Report 2013*. Canberra: Montreal Process Implementation Group for Australia and National Forest Inventory Steering Committee.

Norton, D. and Reid, N. (2013) *Nature and Farming: Sustaining Native Biodiversity in Agricultural Landscapes*. Collingwood: CSIRO Publishing.

Noss, R. F. (1990) 'Indicators for monitoring biodiversity: a hierarchical approach', *Conservation Biology*, 4, 355–364.

Pearson, C. J. (2007) 'Regenerative, semiclosed systems: a priority for twenty-first-century agriculture', *BioScience*, 57(5), 409–418.

Productivity Commission (2005) *Trends in Australian Agriculture*. Research Paper, Canberra: Productivity Commission.

Rockström, J., Steffen, W., Noone, K., Persson, Å., Chapin, III, F. S., Lambin, E. et al. (2009) 'Planetary boundaries: exploring the safe operating space for humanity', *Ecology and Society*, 14(2), 32. www.ecologyandsociety.org/vol14/iss2/ art32/ accessed on 31 July 2018.

Saunders, D. A., *Hopkins*, A. J. M. and How, R. A. (eds.) (1990) *Australian Ecosystems*: *200 Years* of *Utilization*, *Degradation and Reconstruction*. Chipping Norton: *Surrey Beatty* & Sons for the Ecological Society of *Australia*.

Schirmer, J., Berry, H. L. and O'Brien L. V. (2013) 'Healthier land, healthier farmers: considering the potential of natural resource management as a place-focused farmer health intervention', *Health & Place*, 24, 97–109.

Soils for Life (2016) *Innovations for Regenerative Landscape Management. Case Studies of Regenerative Landscape Management in Practice*, 3rd ed. Manuka: Outcomes Australia.

Thackway, R. (2013) 'Applying a system for tracking the changes in vegetation condition to Australia's forests', in Brown, A. G., Wells, K. F., Parsons, M. and Kerruish, C.M. (eds.), *Managing our Forests into the 21st Century*. Proceedings of national conference, 4–7 April 2013, Canberra: Institute of Foresters of Australia, pp. 79–91.

Thackway, R. (2014) *VAST-2 Tracking the Transformation of Vegetated Landscapes, Handbook for Recording Site-based Effects of Land Use and Land Management Practices on the Condition of Native Plant Communities*. Version 3.0, June 2014, Brisbane: Australian Centre for Ecological Analysis and Synthesis, Terrestrial Ecosystem Research Network, University of Queensland.

Thackway, R. and Freudenberger, D. (2016) 'Accounting for the drivers that degrade and restore landscape functions in Australia', *Land 5*(4), 40, 1–20, s1–s15.

Thackway, R. and Specht, A. (2015) 'Synthesising the effects of land use on natural and managed landscapes', *Science of the Total Environment*, 526, 136–152.

White, C. (2012) 'Pasture cropping: a regenerative idea from down under', *Acres USA*, 42(7), 1–5. www.pasturecropping.com/images/PDF/Acres%20USA.pdf accessed on 31 July 2018.

Yapp, G. A. and Thackway, R. (2015) 'Responding to change – criteria and indicators for managing the transformation of vegetated landscapes to maintain or restore ecosystem diversity', in Blanco J. A. (ed.), *Biodiversity in Ecosystems – Linking Structure and Function*. www.intechopen.com/books/biodiversity-in-ecosystems-linking-structure-and-function/responding-to-change-criteria-and-indicators-for-managing-the-transformation-of-vegetated-landscapes accessed on 31 July 2018.

Yapp, G., Walker, J. and Thackway, R. (2010) 'Linking vegetation type and condition to ecosystem goods and services', *Ecological Complexity*, 7(3), 292–301.

13

Transcoherence

Labels and wicked problems

Craig A. Ashhurst

Introduction

'A __ by any other name would smell as sweet.'

This chapter is a story about working through a dilemma about labels. I want to get rid of a bad label, 'common sense', and introduce a new label, 'transcoherence'. A lot of people are likely to have a problem with my doing this because labels are important to people, they have inertia and are hard to change. Given this, why not just use the current labels? To answer this leads me to a long-running argument about the relationship between thinking, values and language.

At one extreme, some authors have claimed the language of a society determines the way people think, their culture and how they perceive the world. Labels then can be used to force people to think in certain ways, a form of linguistic determinism (Hussein, 2012). Changing the term 'refugee' to 'illegal immigrant' might be an example. The belief then would be that the new label will result in a change in how people in those groups are perceived and this would lead to their being treated differently.

At the other extreme is the belief that thought is autonomous, little influenced by language (Blackburn, 2016). People with this view may look at the label change and say: 'It's just semantics. Don't worry about it.'

In the middle lie those who believe that thought and values influence language and, at the same time, language influences thinking and culture (Lucy, 2015; Whorf, 1959). Where someone sits on this spectrum can significantly affect how they view the idea of introducing new labels.

Most of us don't think a lot about words, grammar or language. We just communicate using what seems natural to us until something jars our sensibilities. It may be an unknown word, a strange phrase or the 'wrong' use of a metaphor. Something that pulls us up short and makes us think about the word that was just used. Something like 'transcoherence'.

Let me do a simple test. What word would you put in the blank space in the quote at the beginning of the chapter? For many people reading this, it will be obvious, to the point of wondering why I would even bother asking. They might even say that the answer is just 'common sense', that only one words fits, 'rose'. Of course the quotation reads 'a rose by any other name would smell as sweet.' Rose is the right label!

Yet if I fill in the blank with the word 'cake', what does that do? Is it jarring? Why? Surely I can use any word that makes sense to fill in the blank? Many would say that I could not or should not use cake because it is the 'wrong' word: that my suggestion is incoherent. If you assumed that the word in the blank can *only* be 'rose', you did so not because of common sense, but because you share a narrow history with the original author and those of us who are familiar with his work, the play *Romeo and Juliet* (Shakespeare, 1597: Act 2, Scene 2). Common sense only works for those who hold the sense in common. If someone belongs to a social world in which Shakespeare's play is widely known, filling in the blank with rose is natural. On the other hand, if someone comes from a different cultural world, maybe cake would be seen as a reasonable word to use. My reason for dragging Shakespeare into this introduction is because the focus of this chapter is about the problem of labels and Juliet was up to her eyeballs in label problems.

Getting rid of a bad label

Let me look a little more closely at getting rid of a label. The quote I have used from Shakespeare finds Juliet wanting to get rid of Romeo's label. Thinking out loud about him, she says: 'Tis but thy name that is my enemy.' His family name is Montague and her family is at war with his. But she loves him. So she wonders about solving her problem by getting rid of his label. Or, as she puts it: 'O, be some other name!'

In my case, the label I want to get rid of is 'common sense'; not because of a family war but because of the damage associated with its use.

Part of my work as a management consultant is on risk. In this space, I frequently hear people talk about common sense (Long, 2012). It is a label thrown at others, usually in a negative way. Someone 'doesn't have any common sense', 'they should have just used common sense' or 'anyone with common sense would have …'. Common sense is supposedly the knowledge that we all have, or should have, and only 'idiots' don't use it.

I want to get rid of this term because it wrongly labels the person it is thrown at, and this can create confusion, conflict and anger. There is no self-evident norm of knowledge that can be somehow magically drawn on by everyone. In every group I work with, some people know things that others don't. Some things are common knowledge in particular occupations but are completely unknown to those who don't work in that area. To not know something does not make someone an idiot or morally at fault, but that is what the belief in the myth of common sense leads to.

To go further, belief in common sense can destroy *teams of diverse people working together to tackle complex social issues*. Whether they are mixed teams of experts within an organization or community groups formed to tackle a local concern, these groups frequently fracture along fault lines of difference; breaking into conflicting subgroups or individuals, all colliding with one another. These collisions in turn have a detrimental overall effect on team-level outcomes (Meyer et al., 2014), including impacts on group processes, communication, decision making, performance, satisfaction and cohesion, resulting in team fragmentation and dissolution (Adair, Liang and Hideg, 2017; Thatcher and Patel, 2012).

In my experience, this doesn't usually occur due to malicious intent, but rather as a result of the mistaken belief that we share a 'common sense'; that we all 'live in the same world'. Consequently, when someone in the group behaves outside our expectations, we evaluate their actions by our own 'common sense', leading us to the conclusion that they are 'stupid', 'evil' or 'bloody minded'.

Yet diverse groups don't have to be like this. Even in everyday language we hear comments such as 'What world are you from?' or 'You could never do that in the real world!' Perhaps these comments are closer to the truth than the speakers realize, since they convey a sense that the person being addressed is functioning in some form of alternative reality that is nonsense to the observer.

In fact, disagreements between diverse team members often do reflect a deeper conflict of colliding trajectories of whole, alternative, but still coherent views of reality. As one author has said:

> [I]t becomes increasingly clear that people live not in the same world with different labels attached to it but in radically different conceptual worlds (Hiebert, 2008: 15).

Teams tackling complex social problems are often filled with people from very different worlds and being aware of this is an essential first step to them being able to work together successfully. In the same way, Juliet had to deal with the reality that Romeo was from another world. She was embroiled in a wicked problem and by itself a simple change of label would not solve that problem.

'Wicked problem': an example of a successful new label

Changing a label is not necessarily a waste of time. It can encourage people to think differently and more clearly. One example is the term 'wicked problem'. It's one of a number of labels that came about because of the rising complexity of social problems that could not be solved by technical experts. These problems involved groups of people from many different social worlds, who often shared little 'sense' in common. As a new category of problem, a new approach was required to tackling them (Basadur, Conklin and VanPatter, 2007: 3). This resulted in the generation of a number of new labels, such as:

- wicked vs. tame (Rittel and Webber, 1973)
- messes (Ackoff, 1974)
- ill-structured problems (Mitroff and Mason, 1980)
- policy controversy vs. policy disagreement (Schon, 1999 3–4)
- complex vs. simple (Basadur et al., 2007 : 21)
- social messes (Horn and Weber, 2007)
- messy policy problems (Ney, 2009: 4–11)
- tangled problems (Dawes, Cresswell and Pardo, 2009).

This list can be seen as a collection of young labels each vying for attention and adoption. The label someone chooses from this list reflects her underlying conceptual concern and values. Some labels focus on these types of problem as a social issue, some in terms of policy, others pick up on complexity and a few emphasize mess and entanglement. Yet out of all of the words in the list, the term 'wicked' has come to be the one most authors now use. Originally coined by Rittel and Webber (1973), 'wicked' describes a category of problem that is different from simple, technical, disciplinary or 'tame' problems. But why has this label become the norm?

I can't provide a definitive answer. However, I have run hundreds of workshops on wicked problems and a few crucial reasons keep arising. First, no one has come up with a better label that everyone else will agree to. Second, there is now 'sunk cost' in the label, the idea that people have become invested in something that biases them towards that thing (Kahneman and Tversky, 1979). A third reason is 'strategic ambiguity' (Jarzabkowski, Sillince and Shaw, 2010), where the range of meaning allows the label to act as an 'umbrella term' (Hirsch and Levin, 1999) under which diverse views can coexist.

The introduction of the term 'wicked problem' encouraged people to conceptualize complex social issues in a different way. It is a new label for a new concept. Considering how these problems should be dealt with also gave rise to new labels for new teams that were put together to tackle these problems.

Transdisciplinary teams for tackling wicked problems

Who should work on a problem? It depends on the type of problem. For technical or tame problems, it makes sense to get a technical expert from the relevant area. This doesn't work for our new type of problem. Wicked problems encompass multiple areas of expertise and so new types of individual and team have been required to work together to adequately address them. Over time these have been variously labelled multi-, inter- and trans-disciplinary teams. Each of these terms is an expression of an underlying attempt to involve and integrate the perspectives of multiple experts, to create creative synergies that produce new insights into how to more effectively tackle a wicked problem.

While these are all useful labels, they also create boundaries that resist change and reinforce the orthodoxy that becomes associated with the label. Whatever form of disciplinary team is created, it remains predominantly just that, disciplinary, and usually associated with academic knowledge structures. This means that, although it has become largely recognized that all stakeholders, regardless of qualification, should be involved in tackling wicked problems, the labels themselves tend to constrain the choice of team members to those who have recognized academic qualifications. Many voices are thus left out of the group. Others may be 'consulted' as a 'stakeholder', but they do not sit with the team making the decisions or examining the problem. The teams are diverse but still exclusive. If you don't sit at the table, you don't have a voice.

So what label will encourage all the relevant voices to be heard? I think this calls for another new label.

New labels for new worlds

I am not alone in wanting to broaden the notion of who should be in a diverse team tackling wicked problems. If I want to replace the term 'disciplinary', there are actually quite a few options, each with a different viewpoint on which groups should be included. Psychologists might choose labels based on personality, identity, schema or mental models. Sociologists could pick standard social groupings such as gender, race, class, age, height or education. Linguists may gravitate to discourses, narratives, epistéme or semiotic systems. Scientists might acknowledge different paradigms or research teams that include more than academics. Philosophers could go wide and speak of world views, national perspectives or underlying philosophies. In a similar vein, labels are based in belief or thought systems such as life views, conceptual frameworks and ideologies. Policy and organizational authors lean toward frames, organizational culture or areas of expertise. And these are just the common ones.

Some labels are more specialized and arcane. Bourdieu (1977) introduced a collection of terms – habitus, field and capital – to describe the complex ways in which groups interact and humanity divides. Howard Gardner (1983) invented the term 'multiple intelligences' and, more recently, one of my fellow authors put forward 'multiple knowledge cultures' (Brown, 2008). Both of these last two allow for an individual to have a profile made up of more than one intelligence or culture. Collectives may then form around profile similarities.

Picking a relevant label for a team is not simple. Let me return to Juliet. I don't think she would see a multidisciplinary research team as being very helpful. She would also struggle to form a group to help her using the list of possible groupings from the previous section. The groups connected to her problem might have some form of sociological difference, but hers was primarily a matter of extended families and culture.

Leaving the world of fiction, let's assume I want to have a group of experts working on a current wicked problem. The mass movement of people, for example, is a common and complex problem over much of the world, including my country, Australia. Who should be included in a team created to tackle this problem?

A simple starting point might be to clarify who these people are and what we should call them. The label chosen would inform who should be on a team trying to solve the problem. Until recently (2013), the Australian government officially called these kinds of people 'refugees' or 'asylum seekers'. These labels reflect our international legal obligations and assumptions about why they are coming into the country. Using these labels brings humanitarian groups and groupings to mind.

However, in 2013 a newly elected federal government wanted to change how people think about these people, including the public servants tasked with working on the problem. To direct the thinking of these groups, they mandated a change in the official labels that would be used by government. At the time I was conducting research for my PhD (Ashhurst, in preparation) and was facilitating a diverse group of experts not too dissimilar to what I am imagining here. At one meeting, a team member entered the room and, after saying hello, burst into tears, saying: 'I have to call my refugees "illegals".' This new label brings judicial groups to mind and implies that these people have done something wrong or illegal.

The change in label was intended to change how we thought of these people and consequently who should sit at the table to work on the problem. Now, humanitarian groups might be considered unnecessary and be replaced with police or defence personnel.

Many other examples of wicked problems exist, each having its own collection of groups that have an interest in the problem. All of the labels listed above recognize diversity in group membership by dividing humanity into different coherent collectives of people who share something important. Whatever label is chosen will determine which voices are most likely to be included in the diverse team. While there is value in all of the terms, none of them will safely provide shelter for them all. A bigger umbrella label is needed. For such a label to be useful, it needs to be flexible enough to incorporate the many different configurations of groups that might need to be involved. I have invented a couple of new labels that I think will meet that need: 'collective coherence, and 'transcoherent groups'.

Collective coherence

In using the term 'collective', I am not alluding to the collectives associated with the USSR from the Cold War era, with their Marxist overtones (White, 1986). I am using 'collective' here in contradistinction to 'individual', where the nature of the whole is more than and different from the sum of its parts.

For my new label, I am particularly interested in collectives that are 'coherent'. By this, I mean a group formed when individuals cohere around common elements that are internally consistent and make sense to the members. Linked to this is another assumption, that humans prefer to cohere/collect with others who make sense of the world in a similar way and they identify with these coherent groups (Tsoukalas, 2007). This then creates a boundary between those who are in the collective and those who are outside.

Taken together, the two words 'collective' and 'coherence' generate a label that has an immediately recognizable broad meaning that can be filled out with deeper, more technical meanings as required. All the potential groupings listed above can fit comfortably under the umbrella of collective coherence. By using this label, I am deliberately not privileging any specialized grouping over another and I am opening it up to other types of group not normally included. Juliet's extended family is a coherent collective and even sporting teams or dog breeders fit the definition.

My own PhD research identified specific types of collective coherence underlying conflicts in a diverse team of experts. Collisions occurred at different scales. At a micro level, there were conflicts over ideas and psychologically based personality clashes. At the meso level, there existed incompatible organizational cultures and 'scientifically' grounded opposition between different paradigms. Finally, at the macro level, there were disputes between competing fields of knowledge and interest.

In all cases, the different collective coherences acted as a lens through which a group viewed and made sense of reality. It was used for evaluation, decision making and action. The differences between members of these collectives led to conflicts. The question of how to deal with these conflicts between collective coherences led me to my second umbrella term, 'transcoherence'.

Transcoherence

If tackling wicked problems requires teams with many voices represented, then there is a high probability that conflict will arise along fault lines of difference between team members (Lau and Murnighan, 1998). If conflict in a diverse team is a result of different collective coherences and not just individual disputes, then dealing with that conflict requires more than just standard team building or conflict resolution. I stated earlier that, while transdisciplinary was a useful term, it excluded non-disciplinary collective coherences. My new umbrella term 'transcoherence' utilizes the same prefix, 'trans', but replaces disciplinary with coherence. 'Trans' has been used in multiple ways and I want to draw on many of them. For my purposes, transcoherence is about how people can understand and manage the multiple worlds existing within a diverse team. It has two parts (see Figure 13.1):

a reduction or deactivation of fault lines
an increase in synergy.

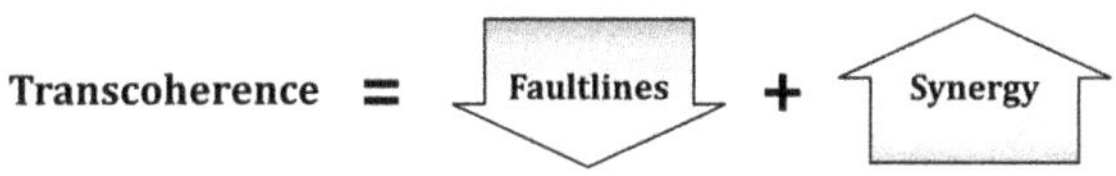

Figure 13.1
A transcoherence equation (adapted from Klein, 2013, p. 4)

A transcoherent approach to team interaction places the emphasis on the relationships between group members and not just a person's individual expertise. This includes both parts of my equation. Fault lines are reduced in strength or deactivated and boundaries are transgressed or bridged to create synergies between team members.

For a person or group to be capable in this way requires a particular mindset and approach to team diversity. Weakening fault lines requires transcending differences and transforming old ways of thinking. Communicating between different collective coherences may well require translation in order to transfer ideas and have them transfuse throughout the team.

For synergies to develop, each collective coherence has to be respected and the different forms of thinking allowed to become more transparent. This enables the group to cross collective boundaries where necessary and for ideas to be transplanted from a single collective to the group as a whole.

My final use of trans is to say that a transcoherent group requires transcoherent or T-shaped experts. This is a label from the management literature (Smathers, 2014; Uhlenbrook and de Jong, 2012). Think of a capital 'T': the vertical bar denotes the depth of one's disciplinary knowledge, while the crossbar represents one's ability to operate across disciplinary, functional or organizational boundaries (see Figure 13.2). Remembering that T-shaped professionals are not be confused with generalists (known as O-shaped). Rather:

> The ideal T-shape is a top expert in one field but he or she can build bridges to other disciplines and is able

Figure 13.2
A T-shaped expert

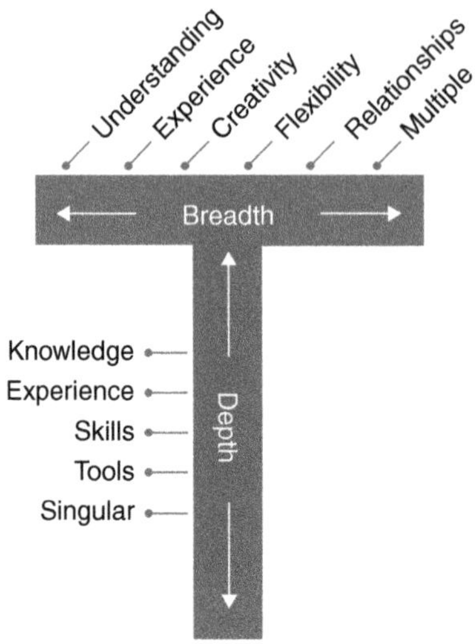

> to think outside of the box. A generalist's profile is characterized by a general knowledge of a wide range of disciplines (each to varying extent), but not by an in-depth understanding of one discipline (Uhlenbrook and de Jong, 2012: 3478).

What should be included in the crossbar of the 'T'? This varies depending on the author, although it often includes characteristics beyond just knowledge, including personal values, organizing, understanding and influencing (McIntosh and Taylor, 2013). Others have described the crossbar as relating to abilities in dealing with multiple people, business functions, systems and cultures (Demirkan and Spohrer, 2015). What is crucial for me is that the crossbar acknowledges the need for capabilities in communicating and working between different collective coherences or worlds.

A T-shaped expert then may belong to any collective coherence and have deep knowledge, experience and skills in that single area. At the same time, he will be able to think and act beyond that single coherence. Their trans or breadth capabilities enable them to have a mind of their own. A team of diverse T-shaped experts should be better able to tackle the collisions of multiple worlds as found in wicked problems.

Implications and conclusion

New labels can be helpful when trying to deal with new types of problem. Wicked problems require a response that includes all the relevant voices. Any form of disciplinary-based team is too exclusive for tackling this type of problem. I have put forward the idea and label of a transcoherent team as an umbrella term that can encompass the many other labels that seek to identify the diversity inherent in humanity.

Labels may not control how we think, but they can influence it. Introducing 'transcoherent teams' as a new label for groups tackling wicked problems allows inclusion of more different and relevant coherent groups when deciding who should have a seat at the table for dealing with the problem.

Suggesting that transcoherent team members be T-shaped shifts the focus of expertise from a singular dimension of deep knowledge to multiple dimensions of expertise. A multidimensional T-shaped expert will recognize that humans are themselves multidimensional, having different ways of understanding their world and relating to others in a rich, complex manner. Yet new labels do not guarantee new thinking or approaches to tackling problems. More is required and many of the other chapters in this book investigate this need and offer new ways to support multiple ways of thinking beyond just changing labels.

Let me finish by returning to Shakespeare using my new labels.

Juliet was faced with a wicked problem. She had fallen in love with someone from another collective coherence that was at war with the one to which she belonged. Thinking through her dilemma, she wonders if his label is the problem, musing, 'Romeo, doff thy name' as a potential solution. When she asks him, 'Art thou not Romeo and a Montague?', Romeo responds with: 'Neither, fair saint, if either thee dislike.'

So what *is* in a name?

I'll leave the last word to our troubled young lovers, in an unlikely modern-day dialogue using my new labels:

> Juliet: I'm sorry Romeo but we are in such a wicked problem that I think the differences between our collective coherences are more than can be bridged by any form of transcoherence. I believe our collectives are on a tragic collision course and so the only way out is to pretend to kill ourselves.
>
> Romeo: Okay, I'm in! After all, what could possibly go wrong?

Acknowledgements

Writing for me is primarily an exercise in dialogue. Therefore, I would like to acknowledge all those willing to talk with me about my ideas and my written work. To Val and John for their sterling work as editors. To the rest of the members of the writing group, a wonderful mixed bunch of people who all support me in their own ways. Dr Liz Clarke and Dr Robert Long who have worked with me for decades and helped me work through many issues. Finally, and most importantly, my wife Pip, the one who has been with me the longest, my greatest critic and most supportive friend. She has listened to every idea, no matter how crazy and clarified my thinking, like no other.

References

Ackoff, R. (1974) *Redesigning the Future: A Systems Approach to Societal Problems.* New York: John Wiley & Sons.

Adair, W. L., Liang, L. H. and Hideg, I. (2017) 'Buffering Against the Detrimental Effects of Demographic Faultlines: The Curious Case of Intragroup Conflict in Small Work Groups', *Negotiation and Conflict Management Research*, 10(1).

Ashhurst, C. (In preparation) *One Team: Where Worlds Collide. An Exploration Into The Development Of Transcoherence.* PhD, Australian National University, Canberra.

Basadur, M., Conklin, J. and VanPatter, G. (2007) 'Rethinking Wicked Problems: Part 1 Unpacking Paradigms, Bridging Universes', *NextD Journal*, (10.1).

Blackburn, S. (2016) 'Sapir-Whorf hypothesis', in S. Blackburn (ed.), *The Oxford Dictionary of Philosophy*. Oxford: Oxford University Press.

Bourdieu, P. (1977) *Outline of a Theory of Practice*. Cambridge: Cambridge University Press.

Brown, V. A. (2008) *Leonardo's Vision: A Guide to Collective Thinking and Action.* Rotterdam: Sense Publishers.

Dawes, S. S., Cresswell, A. M. and Pardo, T. A. (2009). 'From "Need to Know" to "Need to Share': Tangled Problems, Information Boundaries, and the Building of Public Sector Knowledge Networks', *Public Administration Review*, 69(3), 392–402.

Demirkan, H., and Spohrer, J. (2015). 'T-Shaped Innovators: Identifying the Right Talent to Support Service Innovation', *Research-Technology Management*, 58(5), 12–15.

Gardner, H. (1983) *Frames of Mind: The Theory of Multiple Intelligences*. New York: Basic Books.

Hiebert, P. G. (2008) *Transforming Worldviews: An Anthropological Understanding of How People Change*. Grand Rapids: Baker Academic.

Hirsch, P. M. and Levin, D. Z. (1999) 'Umbrella Advocates Versus Validity Police: A Life-Cycle Model', *Organization Science*, 10(2).

Horn, R. E. and Weber, R. P. (2007. 'New Tools For Resolving Wicked Problems: Mess Mapping and Resolution Mapping Processes'. Retrieved from www.strategykinetics.com/New_Tools_For_Resolving_Wicked_Problems.pdf

Hussein, B. A. S. (2012) 'The Sapir-Whorf Hypothesis Today', *Theory and Practice in Language Studies*, 2(3).

Jarzabkowski, P., Sillince, J. A. and Shaw, D. (2010) 'Strategic Ambiguity as a Rhetorical Resource for Enabling Multiple Interests', *Human Relations*, 63(2), 219–248.

Kahneman, D. and Tversky, A. (1979) 'Prospect theory: An Analysis of Decision Under Risk', *Econometrica*, 47, 263–291.

Klein, G. (2013) *Seeing What Others Don't: The Remarkable Ways we Gain Insight*. New York: Perseus Books Group.

Lau, D. and Murnighan, J. K. (1998) 'Demographic Diversity and Faultlines: The Compositional Dynamics of Organizational Groups', *The Academy of Management Review*, 23(2).

Long, R. (2012) *Risk Makes Sense*. Canberra: Scotoma Press.

Lucy, J. A. (2015) 'Sapir-Whorf Hypothesis' in J. D. Wright (ed.), *International Encyclopedia of the Social & Behavioral Sciences* (2nd ed.). London: Elsevier.

McIntosh, B. S. and Taylor, A. (2013) 'Developing T-Shaped Water Professionals: Building Capacity in Collaboration, Learning, and Leadership to Drive Innovation', *Journal of Contemporary Water Research & Education*, 150.

Meyer, B., Glenz, A., Antino, M., Rico, R. and González-Romá, V. (2014) 'Faultlines and Subgroups: A Meta-Review and Measurement Guide', *Small Group Research*, 45(6), 633–670.

Mitroff, I. and Mason, R. (1980) 'Structuring Ill-Structured Policy Issues: Further Explorations in a Methodology for Messy Problems', *Strategic Management Journal*, 1, 331–342.

Ney, S. (2009) *Resolving Messy Policy Problems : Handling Conflict in Environmental, Transport, Health and Ageing Policy*. London:Earthscan.

Rittel, H. and Webber, M. (1973) 'Dilemmas in a General Theory of Planning', *Policy Sciences*, 4(2), 155–169.

Schon, D. (1999) *Frame Reflection: Toward the Resolution of Intractable Policy Controversies*. New York: Basic Books.

Shakespeare, W. (1597) *Romeo and Juliet. The Complete Works of William Shakespeare*. Retrieved 26 June 2018.

Smathers, R. A. (2014) 'The 21th Century T-shaped Lawyer: Lawyers Must Now Acquire Knowledge in Areas Other than Law to Identify Issues, Understand Concepts, Contribute to Teams and Connect Ideas across Disciplines', *Law Practice*, 40(4).

Thatcher, S. M. and Patel, P. C. (2012) 'Group Faultlines: A Review, Integration, and Guide to Future Research', *Journal of Management*, 38(4), 969–1009.
Tsoukalas, I. (2007) 'Exploring the Microfoundations of Group Consciousness', *Culture & Psychology*, 13(1), 39–81.
Uhlenbrook, S. and de Jong, E. (2012) 'T-shaped Competency Profile for Water Professionals of the Future', *Hydrology and Earth System Sciences*, 16(10), 3475–3483.
White, P. (ed.) (1986) *The Dictionary of Human Geography* (2nd ed.). Oxford: Blackwell.
Whorf, B. L. (1959) *Language Thought and Reality*. New York: John Wiley & Sons.

14

Re-imagining person-centred practice in a person-first organization

Julia H. Wolfson

Why the radical social change in person-centred care matters

One billion people of all ages across the globe – one-eighth of our human community – rely on health care, protection systems and support providers (World Health Organization, 2018). In the USA, the number of people needing long-term services and support is expected to double in coming decades, largely due to an ageing baby boomer population (Friedman et al., 2018). In Australia, one in 20 people depends on care (Australian Network on Disability, 2018). The need may be temporary, longer term or lifelong. For many people in these systems, the experience doesn't match their own life priorities and needs for well-being. Too often the person feels unseen and for those in residential or day programmes, day-wasting activities are often the norm (Barol, 2001; CQL, 2017; Wolfson, 2017).

The helping professions have for some time been moving away from a restrictive and imposing custodial-care model to what is commonly known as person-centred care and support. Custodial care means that the system, specialists and practitioners know best what is good *for* you. The focus is on basic subsistence, safety, minimization of risk and maintaining the status quo.

In person-centred care, the person accessing the service is the expert on what makes for a life worth living (Smull and Sanderson, 2009). Despite decades of services under the person-centred banner, the lives of many people dependent on care are constrained by programmes, rules and impersonal routines. These rob the person of opportunities and allow the agency to decide for itself about the person's life.

As a society, we are reeling from revelations of abuse and neglect in environments meant to provide care and support, across the spectrum of health, care, education and religious organizations (Byrne, 2017; Miller, 2016; Wolfson, 2017). Being disconnected from a sense of personal power makes it more difficult to energetically push back when you feel overpowered. Conversely, people in positions of power and influence, who have a low sense of personal power, tend to use their power poorly (Diamond, 2016). This problem is well known, but what can be done about it?

Justice must be done and be seen to be done. However, abuse and neglect cannot just be stomped out. To stop abuse involves creating environments conducive to developing good use of positional power and influence *and* a good sense of personal power. Power can corrupt, but power is also the solution (Diamond, 2016). Positional influence is a familiar aspect of power, but only part of the story. Personal power means to know yourself and have access to your centre when the ground under you is shifting. It is also having the power to detach from the trouble, take a larger perspective, and help move complicated things forward. (Mindell, 1992, 2014). In my book *Applying Deep Democracy in Human Services* (2017), I write comprehensively about an expanded concept of power. I describe a range of innate powers – inner resources available potentially to everyone – that help strengthen the natural immune system for resisting abuse (Wolfson, 2017).

A person reliant on support typically has the least positional power and influence. Paradoxically, this person often has important human lessons to teach about resilience, relationships and community, having endured dehumanizing environments away from the public eye. Formal roles and context determine the extent of positional power. Think of a CEO, a person receiving care, a mother, a funding decision maker. Personal power, on the other hand, is beyond context. You can take it with you wherever you go.

A reimagining of person-centred practice focuses on developing these inner powers among everyone involved (Wolfson, 2017). The goal is twofold: greater personal and collective fulfilment, while stimulating the natural immune system for resisting violence. The capability, for instance, to follow the person's lead with inner curiosity: to ask, to learn and to follow. The power of self-attunement with awareness of safety signals. The power to make conflict fruitful and find meaning in the mess. Developing inner strength from adversity. Discovering and pursuing your own inner direction and path. The inner grit to face injustices of the past with the attitude 'yesterday you, tomorrow me'. The ability to self-renew. The self-belief that comes from wrestling with inner critics and turning them into allies. The skills to facilitate collaboration – in your inner team, the person's team and between various stakeholder groups and perspectives.

This chapter offers a brief overview of a vision of person-centred practice, and what these ideas can look like in an organizational context. I then focus on three person-centred pillars in a power-savvy culture: 1) inner direction 2) dynamic learning hubs 3) engagement facilitation. Before diving in, I'll introduce our team.

Who are we – our inspirations and aspirations

In recent years, in my collaboration with Dr Nader Shabahangi, we are bringing together the fruits of our complementary journeys in elder care, disability, education and health care. Working together in synergy is more fun than working alone. We are each discovering and experimenting with new ideas.

A common approach to teaching staff is to provide regular training. The problem with many training initiatives is the Monday morning effect; that is, what happens in the training room stays in the training room and has little transferable impact on the everyday lives of people served. We found that showing various dynamics in the system in performance mode, including taboo topics and complicated interactions as they surface, can shift mental mountains. We do less didactic teaching and use a 'more show' and 'less tell' approach.

Nader and I use process work in our approach. Process work and its deep democracy philosophy (Mindell, 1992, 1995, 2008, 2011) was pioneered by Arnold Mindell with Amy Mindell and colleagues around the world.

Process work is an information and social theory framework that can be used in any context wherever it is useful. We all know what democracy means – one voice, one vote and freedom of public discourse. But democracy as it is isn't enough to reduce conflict in our world. What's more, there are many who don't make it to the public discourse because they are on the margins, invisible. Arnold Mindell first coined the term 'deep democracy' to describe a welcoming attitude to all stakeholder perspectives and states of mind, including emotions, conflicts and dreams, as the basis for diversity and unity in co-creating new solutions together.

Health, care and support provider organizations tend to attract people from all walks of life, with various gender and sexual identities, economic resources, age groups, health, political and religious affiliations and many cultures and backgrounds – a microcosm of human diversity. Such organizations are natural greenhouses for learning across differences, developing power awareness and improving life conditions for everyone involved.

Process work methods train the awareness to notice, direct from the ground of experience. Everything present in the field of awareness – place, environment, feelings, ideas, hopes and dreams, events, problems, things – makes up an interactive field. What catches your attention comes from your attention – what your mindset will allow in and what it keeps out.

When we work with a person, Nader and I engage with the person's entire experience – the objective reality of their life situation, their subjective feelings and complications, and their deepest dreams. When we work with leaders and teams, we focus on *their* experience. In working with an organization or community, we focus on the experiences of diverse stakeholders and their interactions. Often, contentious issues appear in fleeting signals such as body language, tone, pace and moods and use of third parties. Third parties are views and entities that we do not personally identify with. *Other people have this problem with you, not me.* These ghost-like figures hang in the air like a leaky air conditioner and infect the atmosphere. We see our role as making it easier for the different fragments of information in the system to become visible, interact, be owned and reveal their core essences, to allow for a natural deepening and unifying realizations

Mindell's curiosity to understand the workings of the universe led him first to physics and then to Jungian psychology. He applied theories from physics to psychological and organizational frames of experience: a non-local field effect and the personal–social–global connection. His innerwork approach is a tool for navigating the path between outer and inner experiences. These ideas underpin our approach to person-centred practice:

Field effect relieves the person from the burden of total causality and lessens a pathologizing tendency in a 'you are broken, I am the expert to fix you' attitude. Mindell suggests: 'If your identity changes, the world around you changes as well' (2012, 70). The idea that humans are part of an open, interactive system also humanizes health and support professionals, who belong, like their clients, to the larger social field and universe, and are thus partaking in a similar human journey.

The *personal–social–global connection* interlinks individual and systemic issues. Every personal issue is present in the world, and every global issue in an essence-like way is a personal experience. For instance, we may think of refugees as a world problem, but who among us hasn't at one time felt 'I don't feel at home here', in myself, in the family, in a relationship, at work, or feeling an outsider regarding a particular societal norm. Working at one level impacts on another.

Innerwork. Social issues also exist as inner parts and voices inside each person. Occasionally, you may have enough awareness to notice that the situation in the external environment that disturbs you is also present inside you. Just as the structure of the universe mirrors the structure of the mind, inner and outer worlds are two sides of the same coin. Innerwork is a reflective capacity to facilitate on the inside, complication on the outside, as if the players are inside you. As Buddhist monk Thich Nhat Hanh laments:

> I am the twelve-year-old girl, refugee on a small boat,
> who throws herself into the ocean after being raped by a sea pirate,
> and I am the pirate, my heart not yet capable of seeing and loving. (Nhat Hanh, 2001, 72)

Field effects, the personal–social–global connections and innerwork offer new possibilities to person-centred practice. We are not just addressing the person's needs for their sake, as central as that is. The person's process is also affecting and transforming the entire field of the organization and, by implication, the society that everyone is interacting in.

Follow your inner direction: a person-centred pillar

Life myth

The individual's inner direction is the focus of person-centred practice. The notion of a red thread coursing through the veins of our life is a familiar one (Buber, 1996; Frankl, 2006). Even though we go through many changes over a lifetime, sometimes in hindsight you can see that there's a pattern, a symmetry, a beauty, a consistent contradiction that is uniquely me and repeats itself.

As an analogy, imagine you see a person you haven't seen for many years, walking on the other side of the street. You can't see her face, but even so there is something familiar about her gait, beyond the details, beyond time, something unmistakably 'her'. C. G Jung, the father of transpersonal psychology, introduced the concept of a personal myth to describe essential pattern-like tendencies present throughout life in different ways and settings (Jung, 1963). The word 'myth' indicates the mysterious nature of each of our lives. Our story is both personal and universal (Diamond and Spark Jones, 2004).

Inner direction – an example

Brian, a leader I worked with, mostly comes across as open and friendly. This tendency is pretty close to his usual view of himself. He also has another tendency that is very focused. He is ambitious for people's lives to change for the better and he relies on his more focused part for that. He worries, though, that at times he is too sharp. He gets criticized for overriding people. This disturbs the part of him that wants to be liked. In our coaching sessions, we play with these two parts of Brian's personality, like inner siblings he can get to know, access and use to his advantage. Both exist, although not always in a good relationship with one another. At times, these inner roles jostle for space over the other. The sharp one pushes forward, the friendly one feels crushed. Brian identified his personal growth area as becoming more aware of both parts, so he can be more fluid when one or the other of these inner figures takes over and blocks him.

Brian happens to be an organizational leader, but swinging between the aspects of your inner direction is a universal journey principle for anyone in any role. I am an advocate of integrated learning spaces – people reliant on support and people providing support learning and interacting together as co-learners. The beauty of this mutual learning is that everyone can learn from and with each other in what it means to be human.

The joy principle

The idea of an inner direction deepens person-centred care and support, by taking as our starting point the personal growth and talent development of every person. Rather than create or impose change, we learn to study the changes as they happen and follow them (Mindell, 1993, 74). The only person who truly knows if, when and how, care and services are tuned to their inner direction is the person reliant on support. This person is our guide and inspiration. Her/his experience and interactions are the ground for the support team in co-designing interventions, functions and expanding life-enhancing opportunities that bring happiness. For global facilitator Max Schupbach, happiness means discovering a sense of joy in all areas of life, including those that are difficult (in Kuleshova, 2018).

Inner direction and quality of life

An inner direction cannot be pre-programmed. It is a lifelong current in constant motion, pulling from here to there, at times flowing and at times stuck. How a person's inner tendencies and direction translate into a meaningful and purposeful life, differs from person to person, and even from moment to moment (Wolfson, 2017, 99–121). A quality of life framework gives a reliable basis for discovering what is important to the person, as the focus for designing supports. For one, it might mean in a certain phase, focusing on basic things like food, health, comfort or safety. For another, it might mean getting out more, joining interesting activities and meeting people or reconnecting with family. And someone else might be harbouring a deep wound and want to be heard and known. Most important might be having a home to feel safe and relaxed in. Or feeling welcomed in the community. For many, developing a talent, living a dream, doing something meaningful for others and contributing to a larger cause gives meaning.

Organizational myth

Organizations, too, have a myth, a story, pulsing through the ups and downs, through losing the way and finding the way. The myth feeds the vision: a lode star and an inner motor connected with the question in the world that the organization is setting out to answer. It may be loneliness, care, a future for the planet, making a home, hospitality, safety, good food and enough for everyone.

Being connected with your inner direction makes it easier to know whether your path is connected to the path of the organization. Working simultaneously with personal and professional development is sustenance for people and for the organization. This makes good sense for those reliant on support, for staffing and for financial viability.

Deep democracy coaching

Person-centred practice is about human interaction. For staff and leaders, personal growth is inseparable from skill development and leadership development. The person may be well intended, want to do all the right things and yet be unaware of the impact of her behaviours, her style or actions on others. Deep democracy coaching provides a confidential forum to drop out of everyday duties and pressures. Questions, dreams and disturbances that are often buried have a space in which to surface. The person gets to know various inner tendencies as resources to rely on. With the skilled and goodhearted assistance of the coach, he learns to facilitate interactions in his inner scene. These interactions are mirrored in the outer world and so he becomes more energized and effective in navigating his life and role. The coaching interaction can help shift a victim mindset. From a deep-democracy perspective, disturbances are 'knocking at the door to change the present one-sided identity' (personal communication, Helbling, 2018).

The inner drive for development and joyous teamwork among leaders sets the pattern in motion for a person-first organization.

Overview of a person-first organization

The experience of the person reliant on support is the red thread for guiding the care team and an entire community, for making the shift from a more mechanical and hierarchical view of care. The mechanical approach is based on the expert's view of the person. The role of person-centred helpers – whether paid, voluntary, family members or staff – is to tune in *with* the person we are with, rather than with one's own idea of what is right or wrong *about* the person and their situation. This brings more ambiguity, less prescribed solutions and the need for a deeply felt belief in the abilities, potential and rights of people even if we haven't seen them blossom yet. We stop being convinced that we know best, as there is no longer outer certainty. The inner powers mentioned earlier are resources for living joyously amid contradictions.

A person-first mindset can be applied anywhere it is useful. There is a long list of applications including: planning, care/support issues, case conferences, appointments with specialists, challenging behaviours, new on-boarding, supervising, team meetings, team splitting, paradigm conflicts, tensions and conflict, complaints and grievances, legal issues, licensing visits and compliance reviews, supervision and performance reviews, policy and procedure reviews and documentation updates, problematic email interactions, financial stewardship, reporting and reviews, non-profit board engagement, strategic planning, external partnerships.

From job description to person-first role guide

A person-first role guide is a practical tool for reframing a job description from a list of tasks to purposeful, person-first outcomes. It places the core business of serving people front and centre in everyone's minds and work plans. Performance reviews become learning events for reciprocal exchange. The staff member can use the guide to reflect on results, tensions, questions and learning areas. The supervisor can gain a more accurate view of where staff members are at in terms of their effectiveness and can focus on creative ideas and personal and professional growth areas.

Dynamic learning hubs: a person-centred pillar

Most people do not like being spoon fed bits of information from on high, as a trickle-down solution to information sharing and consultation. Leaders and managers tend to avoid dynamic interaction in large groups for fear of losing control, attacks and fallout. As a result, tensions and criticisms that are not interacted with often escalate into grievances, litigation, staff problems, team splitting, avoidance, work silos, disappointment, burnout and illness (Wolfson, 2017).

Dynamic learning hubs are an interactive space for teams and groups. Role play allows for the whole information system to enter the room. The role of facilitation is to support the various polarities and perspectives to become visible, felt, interact and get to know one another. There is space for doubt and questioning. A learning hub shows rather than tells. The engagement facilitators (see below) work only from their ground of experience. Participants who may typically be unmoved, often free up to new experiences in this open-minded atmosphere (Mindell, 2014).

The inspiration behind the dynamic role play method lies in Mindell's discoveries as a psychotherapist. He found that the therapy room was inadequate for many of his clients, whose problems were connected with larger social and community issues. Anyone can step into any role and get to know different even opposing views. Individual difficulties take on a new light, as something the person carries on behalf of the community and the community can take back it responsibly for aspects of the solution.

Learning hubs welcome fresh and robust exchange. Although there are named facilitators, from a deep-democracy perspective, the moment is the facilitator. Whoever has the awareness at the time, can try to step in. Timing is key. The moment will decide whether an intervention catches alight.

Rather than abstractly talking about an issue, people enter the experience directly. Participants gain a shared understanding of the issues in real time. This can spark unexpected insights and renewed motivation and focus. For example, in one residential team, the group wanted to talk about a hot issue. 'What to do about John, a middle-aged man we support?' 'John eats too much. The doctor has put him on a diet. He steals food. The caring staff were visibly frustrated. Why doesn't he stick to his diet? What if he kills himself from overeating? We feel helpless with John!'

Food is a global issue. It is also common to every place I have worked in: who is allowed to have food, who gets it, when, how much, choice, being denied food. In residential environments, unintentionally, food is often weaponized as a tool of power, punishment and control. While there are at times important health risks to consider, most of us would outright reject such intrusion and surveillance (Wolfson, 2017).

In the above example, it was not long before the interaction began to cycle back and forth. One half of the group insisted that the rules for food intake should be stricter and the other half was pushing back, upset with strictness. I was in the facilitator role, yet I suddenly felt like John, listening and feeling all this. I moved next to Anton, who had assumed the role of John. He stood alone, looking downcast. I spoke as John: 'Be quiet all of you!' I shouted over the confusion: 'I can't stand it! Don't assume you know anything about me! Ask me what is going on for me!'

Everyone stopped in their tracks. A circuit breaker. Anton, who in reality is John's residential manager, stepped out of John's role, spun around to face me (in the role of John), and answered as himself, with a heartfelt tremble. 'I am so sorry for not noticing you. Thank you for waking me up. I want to learn.' The mood in the room changed. There was an overall feeling of relief – a momentary completion. We gathered around to reflect. The team realized they were focusing on food, but bypassing John. Food was the lens for relating to John. No one had ever asked him about his life priorities beyond what was on offer here – the place he was supposed to call 'home'. The team got excited about learning from John about John; his interests, concerns and hopes. The outer issue of food turned into one of inner nourishment.

Such dynamic breakthroughs don't typically result from reading a policy on respect or from being told how to behave. John, in real life, communicates with very few verbal words. Being a quiet person in a verbal world is tough enough. In a service provider world, such a person is often treated as stupid. Everyone at the learning hub had a visceral experience in the shoes of John. Even though he has little talk, John has plenty to communicate!

An open forum is a dynamic learning hub for a whole organization or community, to interact on a topic of broad public interest and concern (Mindell, 2002). We facilitate open forums hosted by residential communities in their local town. Topics have been as varied as sexuality and older people, safety and freedom, where and what is home and welcoming people in altered and extreme emotional states in the neighbourhood. Anyone is welcome to participate.

Learning hubs happen on location, so they can arise spontaneously when the need is felt, rather than waiting for a consultant/external facilitator. A process work learning hub may at times appear chaotic and, at times, deceptively simple. Facilitation is not just a nice little add-on. It is essential for uncovering the deep order underneath the process that contains hidden solutions and directions.

Dynamic learning hubs have a number of advantages, they:

- offer a safe, vigorous interaction space for direct support professionals, health care specialists, teams, family members and community members
- allow for person-centred interactions, even when the person may communicate differently, without words or may not even be present
- make information from the whole system accessible in an inclusive, riveting way from the outset
- depersonalize contentious, unpopular and even painful viewpoints so that no one is scapegoated.

Engagement facilitators: a person-centred pillar

So who does the facilitation? In a person-first organization, everyone is learning to facilitate their inner world and the outer scene as one totality. Who you are as a person, your personality and character, your life experience and the hand you were dealt are natural resources for your unique facilitation style. Facilitation literally means to make things easier. Deep-democracy coaches train facilitation awareness and skills in personal coaching sessions.

Engagement facilitators are people within the organization who are recognized by their leaders and peers as having a natural ability to relate well with others. They help teams to be more related and effective in supporting the inner direction and priorities of the people they serve. They are highly motivated to learn, to facilitate learning hubs and help people join the dots. They bring an open-minded and joyous atmosphere. Even in the midst of difficulties, things are somehow easier to get through in their presence.

Engagement facilitators replace passive use of consultants and external trainers. They ask questions, weave between the various areas, identifying immediate learning needs and coach and teach on the spot. Engagement facilitators are trailblazers.

A process work view on conflict

To transition as an organization to person-centred interactions can be scary. At first, many worry: 'How will people know what's good for them? What if they choose the wrong thing and harm themselves?' Diversity necessarily means difference and with difference comes tension and conflict.

I have found that when tension arises, a shared approach is necessary. Otherwise, the usefulness of disturbances is put on hold. Leaders often reject push-back and interpret unwanted behaviours as attack, manipulation or intentional harm. The self-balancing nature of the system can respond with unpredictable effects.

Engagement facilitators help the self-balancing nature bring more flow into stuck situations by being curious, listening and inviting exchange. They pay attention to facts, to feelings and to essences behind appearances, even in the form of a difficulty or disturbance. Tensions and conflicts are not brushed under the carpet. They are keen to hang in when the going gets tough. They have enough detachment to consider rejection as a need to self-actualize, rather than as a personal affront. Things tend to go easier in their presence.

And never alone. Two at least, so they have each other's backs and can role model the kind of interaction that is a blueprint for a better world. Engagement facilitators learn to follow the direction within what is present, rather than imposing theories and abstract solutions for change. Engagement facilitators are able to be with the pain and discomfort of another person or group. They see strengths rather than helplessness. They make it easier to speak about difficult thoughts, feelings and moods, in a way that helps lighten a heavy or toxic atmosphere. Engagement facilitators help prevent and facilitate conflicts in teams, relationships and the broader organization. The learning path of an engagement facilitator involves:

- being at the centre of their own learning path by following what gives them joy
- becoming expert at noticing a person's strengths and potentials rather than seeing deficits and faults
- developing people on their path to be effective agents of self-powered lives and an organization's vision.

Summary

Mindell's field effect, the personal–social–global connection and the practice of innerwork, stimulate a reimagining of person-centred practice as a dynamic human interaction. Awareness that your inner reality determines how you see and react to the outside is a game changer. Outer and inner reflections are inextricably interconnected in this reimagining.

In a person-first organization, I imagine engagement facilitators walking the line for the inspirations and aspirations of the people receiving support as first principle. I imagine the development of all people being of equal value. I imagine people across the entire community curious and energized, being connected to the inner direction of their lives. I imagine dynamic learning hubs as pop-up spaces for fresh and frank exchange and creative problem solving. I imagine conflict and tension as allies towards new discoveries.

I imagine all this strengthening the natural immune system to resist violence and abuse, in a dynamic field of personal and professional growth and impacting people's lives for the better.

I imagine one billion people counting on us to wake up.

Acknowledgments

The fruits of my life's learning and practice with and from people around the world, morphed onto these pages thanks to my Fenner Writing Group colleagues. In particular, Val Brown and John Harris have supported me with their unstoppable belief in my ideas and work. My collaboration with Nader Shabahangi, founder of AgeSong elder communities, has influenced every aspect of this chapter. I am indebted to Arnold and Amy Mindell, founders of process work. My teachers Max and Ellen Schupbach of the Deep Democracy Institute have given me inspiration, tools and practical applications for my person-first ideas to grow corn; that is, to become actualized in the everyday lives and dreams of people, teams and organizations.

References

Australian Network on Disability, *Disability Statistics*, viewed 6 August 2018, www.and.org.au/pages/disability-statistics.html

Barol, B. (1996) 'The Pennsylvania journal on positive approaches: An overview', *Positive Approaches*, 1(1), 1–3.

Barol, B. (2001) 'Learning from a person's biography: An introduction to the biographical timeline process', *Positive Approaches*, 3(4), 20–30.

Buber, M. (1996) *I and Thou*. New York: Touchstone.

Byrne, G. (2017) 'Prevalence and psychological sequelae of sexual abuse among individuals with an intellectual disability: A review of the recent literature', *Journal of Intellectual Disabilities*, 1–17, viewed 6 August 2018, journals.sagepub.com/

Council on Quality and Leadership (CQL) (2017) *Personal Outcome Measures: Measuring Personal Quality of Life*, 3rd ed. Towson: Council on Quality and Leadership.

Diamond, J. (2016) *Power: A User's Guide*. Santa Fe: BellySong Press.

Diamond, J. and Spark Jones, L. (2004) *A Path Made by Walking: Process Work in Practice*. Portland: Lao Tse Press.

Frankl, V. (2006) *Man's Search for Meaning*, 5th ed. Boston: Beacon Press.

Friedman, C., Caldwell, J., Kennedy, A. R. and Rizzolo, M. C. (2018) 'Aging in place: A national analysis of home and community-based Medicaid services for older adults', *Journal of Disability Policy Studies*, viewed 8 August 2018, journals.sagepub.com/

Jung, C. G. (1963) *Memories, Dreams, Reflections*, 3rd ed. Glasgow: Random House.

Kuleshova, X. (series producer) (2018) TV 21, *Conflictology with Max and Ellen Schupbach*, Russian television programme Psychology 21 – in production.

Miller, J. M. (2016) *The Priests*. Sydney: Finch Publishing.

Mindell, A. (1992) *Leader as a Martial Artist: Techniques and Strategies for Resolving Conflict and Creating Community*. Portland: Lao Tse Press.

Mindell, A. (1993) *The Shaman's Body: A New Shamanism for Transforming Health, Relationships and the Community*. New York: HarperCollins Publishers.

Mindell, A. (1995) *Metaskills: The Spiritual Art of Therapy*. Portland: Lao Tse Press.

Mindell, A. (2002) *The Deep Democracy of Open Forums: Practical Steps to Conflict Prevention and Resolution for the Family, Workplace and World*. Charlottesville: Hampton Roads Publishing Company.

Mindell, A. (2008) 'Bringing deep democracy to life: an awareness paradigm for deepening political dialogue, personal relationships and community interactions', *Psychotherapy and Politics International*, 6, 212–225.

Mindell, A. (2011) *Dreambody: The Body's Role in Revealing the Self*, 2nd ed. Portland: Deep Democracy Exchange.

Mindell, A. (2012) *Quantum Mind: The Edge Between Physics and Psychology*, 2nd ed. Portland: Deep Democracy Exchange.

Mindell, A. (2014) *Sitting in the Fire: Large Group Transformation Using Conflict and Diversity*, 2nd ed. San Francisco: Deep Democracy Exchange.

Nhat Hanh, T. (2001) *Call Me By My True Names: The Collected Poems of Thich Nhat Hanh*. Berkeley: Parallax Press.

Smull, M. and Sanderson, H. (2009) *Essential Lifestyle Planning for Everyone*, 2nd ed. Annapolis: Learning Community.

Toft, K. (series producer) (2017) 'Four Corners', *Fighting the System*, television programme, ABC, 27 March.

Wolfson, J. (2017) *Applying Deep Democracy in Human Services: Diversity, Inclusion and Innate Powers*. San Francisco: Eldership Academy Press.

World Health Organization (WHO) (2018) *Fact Sheet Disability and Health*, viewed 6 August 2018, www.who.int/en/news-room/fact-sheets/detail/disability-and-health

15

Engaging creatively with tension in collaborative research

Rebecca Freeth, Elizabeth A. Clarke and Dena Fam

Introduction

The authors of this chapter are transdisciplinary researchers working in complex, inter- and transdisciplinary projects. Each of us inhabits her own epistemic living space, which means we have our own ways of thinking and making sense of the world (Felt, 2009). This brings together our experiences and understandings from all aspects of our lives. It includes the social and intellectual contexts we are part of, including family, community, profession, organizations and cultures.

In writing this chapter, we have brought these different epistemic living spaces together to form a collaborative authoring partnership, each contributing unique aspects to the paper, but at the same time, forming a whole that is different to the sum of the parts. Our point of connection is an inter- and transdisciplinary research project that combines team members from different ethnic, disciplinary and organizational backgrounds, brought together from around the world with a common purpose of sustainability transformation (Abson et al., 2016).

In this book, Brown, Harris and Waltner-Toews propose that individual agency and independent thinking hold an important key to transformation for just, sustainable futures. They highlight the power of individuals, such as change agents, thinking independently to question and open up closed social knowledge, where assumptions, values, beliefs and understandings are otherwise tacit and unquestioned.

This is also the case in research projects embedded in the academic communities in which each of us works. On the one hand, there is the 'I' of research: individual scholarship and inquiry and, on the other hand, there is the 'we': the collective and collaborative epistemic living spaces that researchers inhabit. We use the term epistemic living space because the diversity of researchers cannot be described only by looking at their disciplinary backgrounds; it also needs to include all aspects of their knowledge and experience, including social, political, physical, institutional and personal.

The idea of interdisciplinarity and complexity have become increasingly entwined (Klein, 2004). Klein and Newell (1997 in Klein, 2004, 2) describe interdisciplinary study as 'a process of answering a question, solving a problem, or addressing a topic that is too broad or complex to be dealt with adequately by a single discipline or profession'. The complex intractable wicked problems of the Anthropocene era are challenging all of us and specialist and 'expert' thinking alone is no longer sufficient.

Traditionally in research, the 'we' has been characterized by relatively closed communities of practice. In recent years, these have been challenged by a growing number of inter- and transdisciplinary endeavours arising in response to the wicked problems of the Anthropocene, where boundaries of communities of practice are breached and there is an increasing need for more heterogeneous epistemic living spaces embracing multiple kinds of knowledge.

There is growing (but not universal) acceptance of the need for a greater diversity of approaches to tackling wicked problems (Australian Public Service Commission,

2007; Head & Alford, 2008; Rittel and Webber, 1973). There is also a need to examine the characteristics of and relationships between, individual thinking and agency ('I') and collective and collaborative approaches ('we'). We are interested in how this relationship can be used to constructively tackle wicked problems and contribute to a just and sustainable future.

In this chapter, we explore the tensions around the 'I' and 'we' paradox and how they emerged within our collaborative research project. Through a series of significant project events, we describe the ways in which these tensions emerged, how we responded to them and the consequences and outcomes of these responses. In particular, we examine how this tension shapes and is shaped by (1) the wickedity of the problem, (2) the level of self-organization and adaptation in our individual scholarship and (3) the degree of heterogeneity of the epistemic living spaces of the various constellations and combinations of the 'we' of team work.

The key events in the project that we draw on include annual all-team retreats and a series of outcome mapping workshops. Our data include participant observation notes and a series of semi-structured interviews with other project team members.

Exploring the paradox of 'I' and 'we' for collective coherence

A paradox is a seemingly contradictory situation that resists simple resolution. A paradox contains two or more elements, like the 'I' and 'we' in interdisciplinary projects, that appear to be mutually exclusive and that appear to operate at the same time (Kahane, 2010). A problem has a solution, whereas Bohm warns 'if the mind treats a paradox as if it were a real problem … the mind is caught in the paradox forever' (Bohm, 2004, 73).

Brown and Harris (2014) emphasize the importance of reframing our ideas of paradox and the tensions they create and the need to look at relationships between opposites, including individuals and society, parts and wholes, stability and chaos and rationality and creativity. They accept with Bateson (1973) that reflection on the patterns that connect is more important than concentrating on the differences that divide. This provides an alternative to the 'binary logic that dominates modern Western thought and the problem of paradox that it creates' and 'fosters the development of creative and innovative mindsets' (Wickson et al., 2006, 1054).

Research teams are increasingly recognizing and engaging with the paradoxes and tensions associated with wicked problems. The particular challenges of these problems require a shift away from the traditional modes of research where individual researchers ('I') work in relatively homogeneous teams and are guided and bounded by a socially constructed knowledge of a community of practice of like-minded scholars who share methodological framings and understandings. In these cases, coherence is relatively easy to achieve. This also applies to multi- and pluri-disciplines, where compatible disciplinary knowledge areas are combined and the emphasis is on juxtaposing, sequencing and coordinating knowledge and approaches (Klein, 2010, 16; Max-Neef, 2005, 6).

By contrast, creating together an inter- or transdisciplinary team brings to light differences in epistemic living spaces, but more particularly in methodological framings and understandings, which immediately escalates tension and can lead to considerable intellectual and emotional discomfort (Smith and Berg, 1987). There are broadly two possible responses to this discomfort. First, by avoiding or dissipating tension through a siloed or 'additive' multi-disciplinary approach, where individuals continue with their scholarship and connect with their disciplinary community, creating a fragmented approach (Roy et al., 2013).

The second possible response is that team members can give the 'I–we' paradox 'serious and sustained attention' (Bohm, 2004) and gain insight into the contradictions and the conditions that create them. In this case, individuals can exercise individual agency through independent creative thinking ('I'), as well as engaging with the interdisciplinary team through collective thinking ('we'), which requires a greater engagement with difference. Here, the emphasis is on integrating, linking, focusing, blending, transcending, transgressing and transforming (Klein, 2010). This latter approach is potentially paradoxical and creates its own tension, as the individual is at the same time acting as a free and independent, self-organizing agent within the project, while at the same time connecting, relating and contributing to the team effort through collective coherence.

Collective coherence infers a sense of unity where two or more different things or ideas are present, and generates the 'confidence of feeling grounded and … that promotes a willingness to act' (Letiche et al., 2011, 392). It is not consensus, but instead infers an approach that embraces diversity of understanding and requires dialogical approaches, pattern recognition and reflexivity. Collective coherence enables a group to develop a shared (but not necessarily homogenous) understanding of reality, as described by Craig Ashhurst in Chapter 13 of this book.

Navigating the 'I' and 'we' tension in a complex inter- and transdisciplinary project

For illustration, we turn to a case study of the collaborative process in a project in which each of the authors has been involved. This project, based at a German university, is a latticework of inter-, trans- and disciplinary research aimed at tackling some of the most intractable challenges of sustainability through institutional restructuring, reconnecting humans and nature and rethinking how we understand and act in the world. The co-located interdisciplinary team comprises 23 individuals from multiple disciplinary, geographical and linguistic 'homes'. Transdisciplinary engagement includes collaboration with partners from outside the academy (Brown, 2010; Klein, 2014; Lang et al., 2012; Thompson Klein, 2004).

We use this example as a way of thinking about enhancing engagement with the 'I–we' paradox in a research context, through our experiences and observations in this project. The diversity of the team and the initial problem framing created challenges as well as opportunities for members of the project team. On the one hand, we were being exposed to new and exciting ideas, visions of sustainability and research methodologies. On the other hand, we struggled with the shift from individual scholarship embedded in disciplinary and multi-/pluri-disciplinary thinking, to the tensions associated with higher levels of individual agency combined with higher levels of team heterogeneity (see Figure 15.1).

While most members of the project are strongly oriented towards interdisciplinarity, this does not necessarily mean that they also have a preference for collaboration. For example, one team member expressed their affinity for interdisciplinarity as 'I don't know anything else.' Later, when describing their orientation to collaboration, the same person stated: 'I like doing my research independently and I like having full ownership over my research.' As Klein (2010) notes: 'Many believe that interdisciplinarity is synonymous with collaboration. It is not.' We believe that individual agency – expressed as independent thinking or taking initiative and leadership – is a necessary ingredient

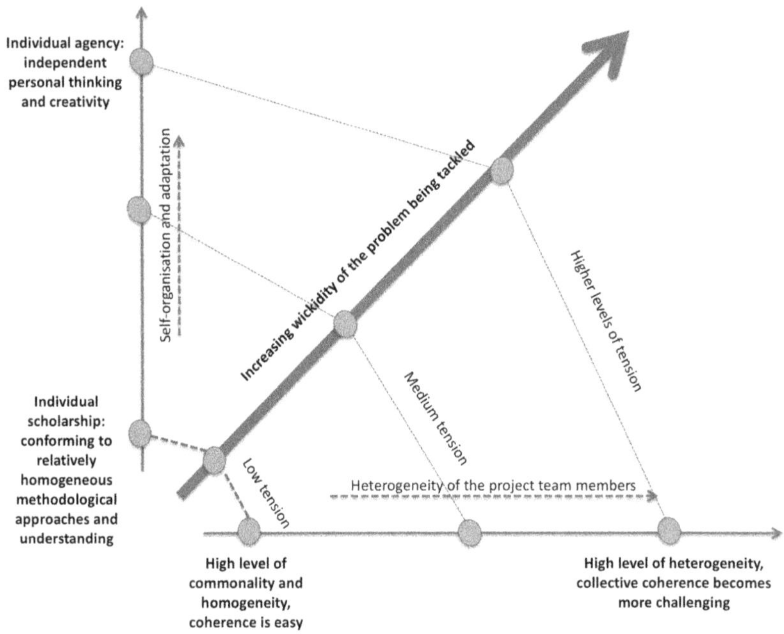

Figure 15.1
Navigating the tensions between the 'I' and the 'we' in research. With increasing wickedness of the problem being researched, the greater the need for individual agency and independent thinking and creativity, along with the need for increasing heterogeneity of the combined epistemic living spaces of the team. This in turn increases the level of tension between individual approaches ('I') and collaboration ('we')

of interdisciplinary research collaborations, as long as the 'I' of the individual researcher remains in a healthy tension with the 'we' of the research team.

Where researchers in the project team have a preference for collaboration, a clearly demarcated difference became evident in terms of whether this is with one other person, a small group or the full project team. In other words, there are multiple constellations of 'we'. Between the 'I' and the full team 'we', it is possible to opt for collaborating with one other person or with a small group of researchers. We have learned not to assume that someone with a strong orientation to collaboration necessarily wants to work in a group, big or small, as the following quote from a colleague indicates: 'It's quite easy to quickly adjust to people's responses and reactions when it's between two people, but doing it with eight people is much more tricky.' Specific criteria usually guide these choices, such as: 'I prefer to work with a small group of people who are dedicating the majority of their effort to that collaboration.'

At different stages of the project, the 'I' and 'we' orientation varied between the low-tension option of individual scholarship and relatively homogeneous backgrounds and contexts of multi-/pluri-disciplinary collaboration, and at the other end, individual agency, creativity and independent thinking and more heterogeneous team engagement and collective thinking mode (Figure 15.1).

To illustrate these shifts and changes in the 'I' and we' orientation and some of the drivers, we focus on a series of significant events in the project, beginning with the start-up phase of the project, in which the project grew from a nascent 'we' of eight principal investigators to the full 23 team members' 'we'. Then we sketch a narrative of three milestone collective events in the project: our annual team retreats. These reveal significant patterns and insights into the team's negotiation of the 'I'–'we' tension.

The eight principal investigators in the team tell the story of collectively designing the project. This was an intensely collective endeavour and, for some, a career highlight because of the intellectual stimulation and trust it generated. A combination of individual agency and collective thinking resulted in a strongly reasoned project document that won funding support and attracted a team of researchers from all over the world. It produced a conceptual framework for the project that seemed, in the words of one of its creators, 'so neat and coherent'.

That early experience of coherence between the principal investigators was disrupted almost as soon as the full team had been recruited. First, the principal investigators and postdocs met to develop a glossary of key terms. This turned out to be a more difficult and incomplete exercise than anticipated, disturbing the sense of epistemic coherence. Then the full team started to interact, which significantly increased the experience of group heterogeneity. As one of the project leaders put it: 'When it was just eight of us, it was quite contained because it was just us dealing with the consequences and we can get on with it. Then we bring in 12 different personalities and you've got how those 12 work with each other and how they each work with their principal investigators, which changes how the principal investigators work with each other.' The dissonance between the desire to create a flat hierarchical structure in the project and the hierarchical nature of academic institutions, meant that it proved difficult for any one person to exercise individual agency in a way that provided direction in these early experiences of divergence. In addition, the multiple institutes in which team members were located administratively created identity questions for the research colleagues, given that each institute has a different focus, methodologies, combination of disciplinary expertise, ways of interacting and reward structures.

Three months after the official project start date, the full team met for the first time for a project retreat and the 'I' and 'we' paradox made a strong appearance. Most of the PhD candidates had just completed their first task of the project, which was to develop their research proposals. Tension manifested between the PhDs' freshly conceptualized individual research project proposals and how their research fitted into the three dimensions of collective research: the place-based case studies (for collective transdisciplinary research work), the research integration process and the project as a whole.

During the course of the retreat, this tension intensified. The locations of the transdisciplinary case study, the topics and partners had not yet been finalized and PhD members of the team were concerned whether the individual research agendas to which they had already committed would fit with the transdisciplinary research agenda. If not, what would this mean for the integration process and would they still be able to contribute to the overall project as envisaged? In a bid to address the growing tension, expressed as rising levels of confusion and frustration, the project leaders met and then announced three decisions. First, that the PhD candidates should prioritize their individual research. They were encouraged to find a way to contribute to the transdisciplinary work, but this would largely be the responsibility of their supervisors. It was acknowledged that making a retrospective fit was not ideal and might not be feasible. Second, that PhDs would not be involved in selecting case studies, on the basis that this would place an unfair degree of responsibility on them. Last, that responsibility for integrating the different work packages and realizing overall project goals was the task of the principal investigators and postdocs, not the PhDs. One of the project leaders stated: 'The most important thing for the PhDs is that you get a PhD out of this project and you don't feel responsible for the overarching project.'

From the perspective of the 'I–we' paradox, these decisions indicated that PhDs should (1) prioritize individual scholarship over collective thinking; and (2) prioritize individual scholarship over individual agency. This also sent a clear signal about diminishing agency for the PhDs in general.

Trying to dissolve the 'I–we' tension in this way had two unintended implications for ongoing navigation of this paradox. First, it created a new tension between a more staid, individual scholarship approach and aspirations to blaze an intense trail of inter- and transdisciplinary collaboration and integration. Second, it initiated a pattern of prioritizing individual thinking early in the project that repeated during the ensuing months and years. This pattern has made collective coherence much more challenging. Figure 15.2 illustrates this strategy of trying to manage tension by retreating to homogeneity and individual scholarship, while trying to tackle a wicked problem, with the resulting increase in tension and pressure to simplify the problem focus.

A year later at the second retreat, there were heated exchanges about the task of research integration, particularly when we discovered that we had divergent ideas about what integration meant, how to do it and whose responsibility it was. Some of us highlighted the process of integration while others highlighted the products of

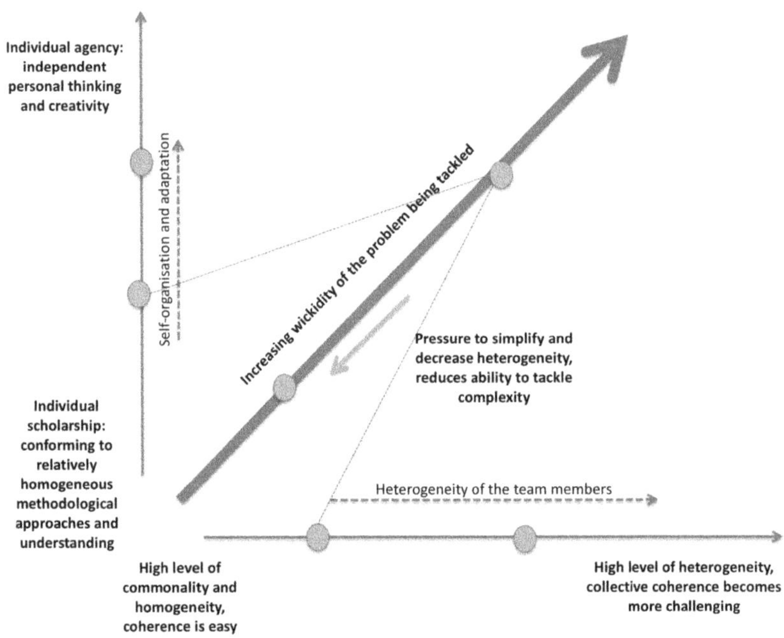

Figure 15.2
The first team retreat: Navigating the tensions between 'I' and 'we'. The tension was diffused and avoided by focusing on individual scholarship. This resulted in increasing pressure to simplify and reduce heterogeneity, making it more difficult to deal with wicked problems

integration, and there was a difference of opinion about whether the integration outcomes should be emergent or more tightly planned and managed.

After the retreat, there was little follow-up on the topic of integration. The tension was avoided. Some members of the team wondered whether integration had been dropped altogether. One of the authors discussed this with a principal investigator, who said that although integration remained a priority, the 'negativity' it seemed to provoke at the retreat had made them reluctant to mention the word at all.

Over the course of the first two retreats, patterns to diffuse or avoid tension were established. However, the paradoxes that caused the tension did not go away. At the start of the third retreat in 2018, the pattern of diffusion and avoidance was repeated when we encountered divergent opinions on key concepts. On each occasion, before tension could build, a principal investigator proposed a way forward that involved bypassing the issue. Both used the word 'avoid' to motivate their proposal. The pattern was well and truly in evidence.

However, in the afternoon session, there was a change. Three boundary objects had been introduced as central concepts for the overall project for the purpose of enhancing collective coherence. These boundary objects are key concepts that are flexible enough to be accessible for a range of epistemic understandings, but at the same time sufficiently immutable to create anchors to connect different understandings (Akkerman and Bakker, 2011; Star and Griesemer, 1989). There had been significant dialogue on these boundary objects in the lead up to the retreat.

The boundary objects in this project related to the concepts of transformation, systems thinking and collective and transdisciplinary research practices. In the afternoon session, there was an extended discussion among small groups about these boundary objects and how they related to their own and their colleagues' work. The resulting discussion became an engaged and open dialogue, opening up the possibilities of engaging the 'I–we' tension more constructively. These discussions allowed individual researchers to engage in a dialogue with their colleagues to share their own vision and ideas for their research around the shared boundary objects. The general feedback about this exercise was very positive with one researcher stating: 'This is the best conversation I've had in the whole project.'

On the second day, with not much time left before closing, there was a facilitated session to identify and plan collective activities in pursuit of research integration. The format was a full-team plenary conversation, sitting in a semi-circle of chairs. The team hit a difficulty fairly early on; some people were ready to dive in and contribute their ideas, while others wanted to understand the parameters of the conversation. The enthusiasm of people with ideas started to flag. Noticing this, the facilitators framed what they saw: 'We are a diverse group; we have different needs and we work at different paces. Some people want to contribute but won't be able to until they fully grasp what they are contributing to. Could those who are ready, but getting frustrated, stay engaged long enough that we're all on board?' In other words, could we stay in the tension? There was agreement to do so and within the hour we had created a satisfying list of ambitious but realistic collective activities, which were then filtered through a series of unanimously approved criteria to produce a prioritized shortlist, with decisions about how to proceed. A week after the retreat, a comprehensive plan was presented back to the full team for discussion, in a meeting that produced high energy, clarity and goodwill.

Despite the diminution of individual agency experienced at the first retreat, subsequent responses of the PhDs differ. Some are exercising individual agency in support of the overall project, realizing its collective goals of meaningful collaboration (process) and deep integration (content). In so doing, they are choosing to remain in the 'I–we' tension, to the benefit of collective coherence. In contrast, some members of the team are exercising individual scholarship with minimal engagement in the collective coherence. Rather than staying in the tension of the apparent contradictions between the 'I' *and* 'we', they see it as a choice between 'I' *or* 'we'. As the final year of the project progresses, it remains to be seen how effectively the project team will be able to continue to constructively and creatively engage in this tension.

Embracing tension in collective thinking through dialogue

Reflecting on the events of the project so far prompts questions about how, when and why the team was able to engage with the tension associated with the paradox of 'I' and 'we' in the project. Introducing dialogical practices (and making time and space for them) has been an important process for strengthening both individual agency and collective thinking. The practices bridged more diverse epistemic living spaces, enabling a greater focus on the ambitious and exciting overall goals of the project. In this section, the authors discuss and analyse the role of dialogue for constructively engaging tension for collective coherence.

Table 15.1 presents four ways of engaging conversationally with paradox in groups, progressing from serial monologue to generative dialogue. The first two ways (serial monologue and engaged monologue) enable expression of individual perspectives. There is an important place

Table 15.1 Four ways of engaging with paradox in collaborative groups

Ways of engaging	*Characterized by*	*High tension results in*	*Implications for tension*	*Implications for paradox*
Serial monologue	Turn taking in telling own perspective; talking in order to be heard Mode: downloading, advocating or presenting	Withdrawal: to reduce tension and avoid conflict Outcome: Return to one's own perspective	Tension sags	Paradox becomes invisible but does not go away
Engaged monologue	Exchange of own perspectives Involves listening Mode: competing or debating	Clash: To be right and prove the other wrong or to defend own view Outcome: hardened different perspectives	Tension increases, possibly to breaking point	Paradox becomes highly charged, which can result in either protracted conflict or avoidance
Reflective dialogue	Exchange to learn about other perspectives Involves inquiry, self-reflexivity and empathy	Engagement: to explore own and others' assumptions and more deeply understand	Tension is held, with potential to be (temporarily) resolved	Paradox becomes more familiar and manageable
Generative dialogue	Emergent exchange with other perspectives and with the bigger picture that created the paradox	Breakthrough: staying with high tension catalyses a paradigmatic shift and sense of collective flow	Tension is transformed into new awareness	Paradox becomes a source of new insight and learning and generates collective coherence

Source: Adapted from Kahane (2008); Scharmer (2008); Ashhurst (personal communication)

for this in collaborative research teams. Each researcher comes to the collaborative process with a creative 'mind of one's own' and valuable experiences of exercising that mind in order to realize a research agenda. The group as a whole benefits from exposure to each person's viewpoint. Individual perspectives tend to be delivered in a presentation format in the academic environment (serial monologue) and are then subject to debate (engaged monologue) so as to strengthen each other's thinking. Individual agency is welcomed as an important resource to the collective. The first two project retreats prioritized this kind of exchange, so that we learned about each other's research interests, methodologies and agenda, and sought to make them more robust.

However, in the context of tackling a wicked problem, serial and engaged monologue is not enough. In collaborative interdisciplinary research, it does not provide a way of navigating the kinds of paradox and tension that such research teams encounter. In a serial monologue, the tension between different methodological approaches and understandings can make team members uncomfortable and this can result in withdrawal, after which the tension becomes slack and the paradox becomes less visible. However, this release from tension for team members is likely to be temporary. In an engaged monologue, the clash of polarized perspectives may create an almost intolerable degree of tension that can render the issues behind the paradox undiscussable (Donovan, 2014).

'Interdisciplinarity is not a monologue' states Klein (1996, 216), reiterating Bakhtin's critique that monologue fails to bring different perspectives into meaningful conversation (Nesari, 2015). This holds us back from deepening our individual and collective understandings of wicked problems. Monologue also closes down opportunities to deepen our appreciation of each other, allowing us to dwell in stereotypes and myths rather than learning from our differences (Baker et al., 2002). This can result in heightened incoherence (Bohm, 2004).

In the academic context, which continues to reward individual scholarship and achievement over collective scholarship (Weingart and Padberg, 2014), there is value in harnessing individual thinking and agency to enrich collective thinking and agency. If we only have individual thinking and agency, this can create a degree of incoherence that becomes unworkable in a team environment.

As an antidote to incoherence, dialogue can enable us, collectively (i.e. 'we'), to examine and address fragmentations, abstractions, assumptions and prejudices incubated in the individual mind (Bohm, 1992, 2004). Moreover, dialogue makes it possible for individual and collective thinking to co-exist, productively activating the 'I–we' tension so that new or renewed collective understanding emerges out of the exchange of individual ideas. Dialogue is 'an intersubjective process whereby an individual maintains a sense of self while at the same time is aware of and open to the influence of others' (Baker et al., 2002, 60). Bohm (1992) claims that dialogue has greater potential than any other methodology to generate 'shared meaning', which does not necessarily imply consensus. Indeed, emergence of greater interpersonal understanding and meaning can develop either by moving towards convergent thinking, or sticking with the frustration of divergent thinking. Interdisciplinarity in collaborative research has potential to be a dialogue, enabling a flow between a mind of one's own and collective coherence.

Returning to Table 15.1, the third (reflective dialogue) and fourth (generative dialogue) ways of engaging with paradox in groups represent constructive approaches to address the tensions inherent in paradox. Reflective dialogue involves curiosity about others' perspectives, with an interest in understanding what makes them different from one's own. The tension is held as the polar ends of the paradox are explored. Strengthening the 'I' paradoxically strengthens the collective. As collective understanding dawns, there may be a sense of temporary resolution of the tension. This is what process work calls a 'cool spot' following a 'hot spot' in group dialogue (Mindell, 2007). Although the contradictions that produce the paradox will probably remain, they become more accessible for ongoing engagement.

The fourth option, generative dialogue, is sometimes possible. In generative dialogue, members of a research team stay engaged with high levels of tension and hence open up windows onto new insight, showing the paradox in a new light and holding potential for deep collective coherence.

As a project team, we have begun to include dialogical ways of engaging with each other about our individual work and the paradoxes we experience in our collective work. In the process, we have started to shift away from patterns of withdrawal and avoidance, and are developing a higher tolerance for collective tension. Learning to navigate the 'I' and 'we' more adroitly has potential to create more synergy between the mind of one's own and the collective mind, which enhances collective coherence.

Nevertheless, despite the proliferation of complex, inter- and transdisciplinary research projects and the associated benefits of individual empowerment and collective coherence, dialogue practices and skills are not yet well established in many research settings. Dialogue processes may be treated as inimical in research settings that prize concrete output above all else. There is rich opportunity in learning to pursue the promise of dialogue in ways that concretely support collective learning outcomes.

Conclusion

The biggest challenge for us as we navigate the wicked problems of the Anthropocene and strive for sustainability transformation is to not only gain the best from each individual but also to invite them into the collective. This quantum effect is possible. To achieve this, each person needs to feel safe, inspired, comfortable, uncomfortable, challenged, appreciated, trusting and trusted.

Our strategy for this is to pursue the promise of dialogue, which can make visible the various paradoxes and accompanying tensions. When such tensions are hidden or avoided, individual agency and collective coherence will be difficult to achieve. Discussing and reflecting on these core issues is not a distraction from the real work of research projects, but should be considered an essential part of process, outputs and learning.

Strong individual agency should mean it is possible to be decisive and to contradict without fear of repercussions and with a sharp edge to rule out complacency. In addition, somewhat paradoxically, being part of a collective mind (as distinct from group think) adds to the richness of collaboration. In this way, we believe the giant leap may be possible and we can see ahead the glimmer of a more just and sustainable future.

As the research project draws towards its conclusion, we look forward to hosting an international conference. At the time of writing, we are exploring ways of using dialogue to broaden engagement in a larger forum about necessary shifts in thinking and approach for transforming towards sustainability.

Acknowledgments

The authors would like to acknowledge and thank the Fenner Writing Group for its ongoing support and for providing a rich space for thinking and writing in. We would also like to thank Emerita Professor Valerie Brown and Dr John Harris for their ideas and enthusiasm and for all their mentoring and support. We would also like to thank Craig Ashhurst for his valuable suggestions and feedback.

References

Abson, D. J., Fischer, J., Leventon, J., Newig, J., Schomerus, T., Vilsmaier, U. et al. (2016) 'Leverage points for sustainability transformation', *Ambio*.

Akkerman, S. F. and Bakker, A. (2011) 'Boundary crossing and boundary objects', *Review of Educational Research*, 81(2), 132–169.

Ashhurst, C. (2019) 'Transcoherence: labels and wicked problems', in Brown, V. A., Harris, J. A. and David Waltner-Toevs (eds.) *Independent Thinking in an Uncertain World: A Mind of One's Own*. London: Routledge.

Australian Public Service Commission, (2007) *Tackling Wicked Problems. A Public Policy Perspective*, Commonwealth of Australia. Available at: www.apsc.gov.au/ publications07/ wickedproblems.pdf.

Baker, A. C., Jensen, P. J. and Kolb, D. (2002) *Conversational Learning: An Experiential Approach To Knowledge Creation*. Westport: Quorum Books.

Bateson, G. (1973) *Steps to an Ecology of Mind: Collected Essays in Anthropology, Psychiatry, Evolution and Epistemology*. St Albans: Paladin.

Bohm, D. (1992) *Thought as a System*. London: Routledge.

Bohm, D. (2004) *On Dialogue*. New York: Routledge Classics.

Brown, V. A. (2010) 'Collective inquiry and its wicked problems', in V. A. Brown, J. A. Harris and J. Y. Russell (eds.), *Tackling Wicked Problems Through the Transdisciplinary Imagination*. London: Earthscan.

Brown, V. A. and Harris, J. A. (2014) *The Human Capacity for Transformational Change: Harnessing the Collective Mind*. London and New York: Routledge.

Donovan, P. J. (2014) 'Leaders behaving badly: Using power to generate undiscussables in action learning sets', *Action Learning: Research and Practice*, 11(2), 179–197.

Felt, U. (2009) 'Introduction: Knowing and Living in Academic Research', in *Knowing and Living in Academic Research: Convergence and Heterogeneity in the European Context*. Prague: Institute of Sociology of the Academy of Sciences in the Czech Republic.

Head, B. and Alford, J. (2008) 'Wicked problems: implications for policy and management', in Australian Political Studies Association Conference.

Kahane, A. (2010) *Power and Love: A Theory and Practice of Social Change*. San Francisco: Berrett-Koehler.

Kahane, A. (2008) 'The potential of talking and the challenge of listening', *The Systems Thinker*, 14.

Klein, J. T. (1996) *Crossing Boundaries: Knowledge, Disciplinarities, and Interdisciplinarities*. Charlottesville: University Press of Virginia.

Klein, J. T. (2004) 'Interdisciplinarity and complexity: an evolving relationship', *E:CO Emergence: Complexity and Organization*, 6(1–2), 2–10.

Klein, J. T. (2010) 'A taxonomy of interdisciplinarity', in R. Frodeman, J. T. Klein and C. Mitcham (eds.), *The Oxford Handbook of Interdisciplinarity*. Oxford: Oxford University Press.

Klein, J. T. (2014) 'Discourses of transdisciplinarity: looking back to the future', *Futures*, 63, 68–74. Available at: linkinghub.elsevier.com/retrieve/pii/ S0016328714001311.

Lang, D. J., Wieck, A., Bergmann, M., Stauffacher, M., Martens, P., Moll, P. et al. (2012) 'Transdisciplinary research in sustainability science: practice, principles and challenges', *Sustainability Science*, 7(Suppl. 1), 25–43. Available at: link.springer.com/10.1007/s11625-011-0149-x [accessed January 19, 2014].

Letiche, H., Lissack, M. and Schultz, R. (2011) *Coherence in the Midst of Complexity: Advances in Social Complexity Theory*. Available at: books.google.ca/books?id= ycYdKQEACAAJ.

Max-Neef, M. (2005) 'Foundations of transdisciplinarity', *Ecological Economics*, 53(1), 5–16.

Mindell, A. (2007) *Worldwork and the Politics of 'Dreaming'*. www.aamindell.net/ [accessed 07.08.2018].

Nesari, A. J. (2015) 'Dialogism versus monologism: a Bakhtinian approach to teaching', *Procedia – Social and Behavioral Sciences*, 205(May), 642–647.

Rittel, H. W. J. and Webber, M. M. (1973) 'Dilemmas in a general theory of planning', *Policy Sciences*, 4(2), 155–169. Available at: www.springerlink.com/index/ 10.1007/BF01405730.

Roy, E. D., Morzillo, A. T., Seijo, F., Reddy, S. M., Rhemtulla, J. M., Milder, J. C. et al. (2013) 'The elusive pursuit of interdisciplinarity at the human–environment interface', *BioScience*, 63(9), 745–753.

Scharmer, C. O. (2008) 'Uncovering the blind spot of leadership', *Leader to Leader*, 3, 52–59.

Smith, K. and Berg, D. (1987) *Paradoxes of Group Life: Understanding Conflict, Paralysis, and Movement in Group Dynamics*. San Francisco: Jossey-Bass.

Star, S. L. and Griesemer, J. R. (1989) 'Institutional ecology, "translations" and boundary objects: amateurs and professionals in Berkeley's Museum of Vertebrate Zoology, 1907–39', *Social Studies of Science*, 19(3), 387–420. Available at: sss.sagepub.com/cgi/ [accessed March 19, 2012].

Thompson Klein, J. (2004) 'Prospects for transdisciplinarity', *Futures*, 36(4), 515–526. Available at: linkinghub.elsevier.com/ retrieve/pii/S0016328703001903 [accessed May 26, 2013].

Weingart, P. and Padberg, B. (2014) 'Interdisciplinarity again … or still?', in *University Experiments in Interdisciplinarity : Obstacles and Opportunities*. Columbia: Columbia University Press.

Wickson, F., Carew, A. and Russell, A. W. (2006) 'Transdisciplinary research: characteristics, quandaries and quality', *Futures*, 38(9), 1046–1059. Available at: linkinghub.elsevier.com/ retrieve/pii/S0016328706000553 [accessed March 2, 2012].

16

Life and change for a regenerative farmer

David S. Marsh

Farming is fundamental to societies that rely on a regular food supply and employ the practice of agriculture to produce food and fibre from the landscape.

Sedentary societies based on agriculture began about 10,000 years ago. Humans like us have been present for perhaps 200,000 years, so agriculture has only been in vogue for one-twentieth of the time that 'modern' humans have existed. To put it another way, there were more than 7000 human generations before agriculture and only 370 since agriculture began. One could say that agriculture is an ongoing experiment.

Before agriculture, hunting and gathering by small bands of people fed a small human population. Over thousands of generations societies learnt to match their population to the available resources. The size of groups reflected the number of people who could survive the harshest weather conditions in the area that a group lived. This meant that for most of the time there was abundant food. To increase population above this level was to threaten the lives of the group if harsh conditions occurred. These bands had learnt through the hard buffetings of time how many people their land could support.

The dawn of agriculture was probably a haphazard affair as people adapted to a very different life and had to match their population to the food supply, no easy feat in a world with a variable climate. However, the great advantage of agriculture was that it made it possible to produce surplus food, gradually leading to an increased population. It allowed time for reflection and for early forms of government and religious institutions to become part of a gradually more complex society. These changes and developments would have taken many generations and thousands of years.

Because of the development of digital communications and the internet, almost all information is now easily available. For some, this allows them to pursue information that may lead to knowledge and understanding. I think that for the great majority, however, this availability of ideas and processes that are the result of building on foregoing thought has made it increasingly hard for people to develop a mind of their own. To wade through the morass of information in some ways removes us from connections with the natural world for which, I believe, we all have an evolutionary deep feeling, even if in this increasingly urban world it has not had the opportunity to be expressed.

I recently read a book by James Rebanks: *The Shepherd's Life*, about a Herdwick (a hardy breed of upland sheep) shepherd from the Lake District in northwest England. His book speaks a lot about the important connections of sheep and shepherds to their land. His family has been farming sheep on the communal summer uplands of the Fells for at least 600 years. This culture was very dear to him and he wanted to clearly state his feelings for the land and the sheep, so that others could feel its value and importance in building a long-lasting community. These farmers and their flocks are part of the land and know it.

James and his wife decided that they would try and spread this story in other ways, not just through the written word via his book and a blog he writes about being a shepherd. So they began asking schools from nearby industrial cities where land connections may have been lost in the mists of time, to come and visit their farm.

So the children came out and James spoke to them about what life is like being a Herdwick shepherd. At the end of the day the children were given about an hour of free time where they could run in the fields and play in the open space, something novel for city kids. In their evaluations of their experiences, all these children said the best thing was running free on the land in the open. To me this was a real-life example of humans' inherent connection to land.

I grew up in a country town in New South Wales, Australia, one of five children in a family where our parents were doctors dedicated to looking after the health of the community. Care was evident in all facets of our life. My father had a passionate interest in native Australian flora and fauna. We all had a great interest in birds and as I grew up I began a bird egg collection. I loved frogs and lizards and spent many hours of my carefree youth trying to catch them so I could look at them. I wasn't much of a reader until I was about 12, but we did have lots of books with pictures of many sorts of bird, fish, tree and plant, shell and seashore life. I have a very good visual memory and the shape of an animal, bird, fish, reptile was enough for me to name it. This ability persists even now as I approach my 70th year.

We often had picnics and barbecues in the bush and as I grew older I went with my father and a couple of his mates on walking and fishing trips to the Snowy Mountains. These trips increased my thirst for the natural world, its beauty and diversity, but also had a technological facet in watching the dams and tunnels of the Snowy Mountains' irrigation and hydroelectric scheme being built.

We went for some walking trips to the dramatic volcanic plugs of the Warrumbungles, the subtropical rainforests of Binna Burra in the Lamington National Park of Queensland and to the spectacular dolerite peaks of Cradle Mountain in Tasmania. So I was extremely fortunate in the experiences that shaped my interests.

I joined the cubs and scouts and loved the outdoors, bushcraft and community responsibility aspects of these two institutions. Thinking of others and doing a good turn were worthy aspirations for a young man.

My interest in agriculture was sparked by spending time with the McMaster family on their property, Glenwood, in the headwaters of the Talbragar River between Coolah and Cassilis in New South Wales. Glenwood is right on the watershed; from the top of the hills to the east of the homestead the western slopes drain westward into the Darling River and out to sea south of Adelaide, and the eastern fall drains to the Tasman Sea.

At this time I had become a hunter and used to try and shoot kangaroos to feed the dogs on the farm. Handling a rifle safely, always being aware of what was behind a target and never shooting over a horizon were lessons in self-discipline and responsibility.

I left school and went jackarooing on the riverine plains in southwest New South Wales, on a sheep stud, and then to a cattle stud on the volcanic plains of western Victoria. These were experiences where I learnt the physical tasks that make a farm work. There was no talk of conservation back then in the late 1960s and we did not have any instruction on the financial management of a farm.

Some of the products we used for jetting and dipping sheep were proved to be toxic in later years due to their tendency for bioaccumulation, as they built up and were stored in fatty tissue. These were the organophosphates and organochlorines the use of which was questioned by writers such as Rachel Carson in her book, *Silent Spring* (1962), when it was shown they were having devastating effects right through food chains. Our methods for handling these chemicals were haphazard and in today's world would have been seen as dangerous. We, unsuspecting of the possible consequences of our carelessness, would probably have scoffed at wearing protective clothing, such was the 'just get on with it' attitude of those now-distant times.

Our family bought a property in 1966 and we employed a manager until 1971, when, as a raw young man of 22 years, I took over the management of Allendale.

Looking back, my version of success then was related to growing a larger yield of meat, wool and grain than my peers. The way I set out to achieve this was along the lines of: 'apply the latest technology and agricultural research, spend a generous amount of capital on agronomy, work hard and profit will ensue'.

This may have been a worthy goal in some respects, it was what most farmers were doing, but really it was based on an economic model of the human/land relationship, what authors such as Charles Massy call the mechanical mind.

Like many, I had some setbacks to my plans, the climate was variable, markets were at times volatile and external factors over which I had no control frequently led to results that did not match my expectations. We were running a self-replacing merino flock for wool and growing a small area of crop with a share farmer. I thought I could do at least as well as the share farmer, and so I bought some second-hand farming plant and tried my hand at growing wheat and oats. This was in the early 1970s when disc ploughs were the primary cultivation implement. Herbicides such as glyphosate and sprayseed were not yet on the market. Sitting on a tractor pulling a disc plough and watching the soil turn over was mesmerizing, it smelled so good and I felt like I was doing something important.

At that time I was ignorant of the effect of ploughing on soil life, I was unaware that a full breakout pass of a plough destroyed 85 per cent of the macropores (the holes in the soil through which water percolates, made by insects, worms and the root channels of plants), in the paddocks I was ploughing. I was thus severely affecting the capacity of the soil to absorb moisture. The reason we thought ploughing was good was that it *conserved* moisture because it killed all the plants in the paddock. But later we found out that the disturbance, in the presence of moisture encouraged a huge blooming of microscopic soil organisms that went to work breaking down organic matter. In one year the land would be ploughed, usually in mid-October before the grasses had set seed. Then it was left fallow until rain had fallen and plants had begun to germinate and grow. This meant scarifying (scratching up with a tyned implement called a scarifier), to break the roots of the emerging plants, which if left to grow, would use nutrients and water that the crop needed. Scarifying also brought on another blooming of tiny soil organisms that caused further breakdown of organic matter. So we were getting rid of nutrients, effectively destroying the soil structure, destroying much of the fungal life of the soil, impeding its capacity to harvest rain and thus increasing the risk of erosion.

Several disastrous erosion events left gutters down the slopes of cultivated paddocks when soils were exposed with no cover and impaired absorption. Somehow, farmers at that time, while not liking these events, seemed to regard it as an acceptable risk.

I was not one who accepted it. I felt responsible, and deep inside I knew my actions had caused the devastation. The embryo of a mind of my own had been born.

The tension I felt opened my mind to think and question what I was doing and feel that it was somehow wrong and that I must search for a gentler, less risky way of farming. In 1982–1983 New South Wales suffered a very severe drought. There was a dry autumn followed by very low rainfall in spring of 1982. It was a short drought and everyone was fully stocked. The government, after pressure from farming groups, announced a policy that once farm grain and hay supplies had been consumed, there would be a 50 per cent subsidy on grain and a transport subsidy for transporting feed and stock, because they had been convinced that this was a case of genuine hardship, caused by exceptional circumstances. The result of this policy was that huge areas were completely denuded, dust storms were common, water sources were failing, depression was widespread in the farming community as farmers could not cope with the devastation they witnessed daily. Farmers felt their financial viability was threatened as markets collapsed and stock were almost valueless.

We were impacted by our farm losing cover, we had reduced sheep numbers but not by nearly enough, in fact most of our thinking was on feeding stock to get them through. The scenes on Allendale were depressing and I knew it was looking so bad due to decisions I had made or failed to make. It did not 'just happen', I was an agent in the outcome, after all, the land between the fences and the road was not bare of cover; it had to be how we were inflexibly running livestock in a variable climate!

This was another tension moment that cracked my mind open and shocked me into vowing to find a different course so that this would never happen again. I cared, but that was not enough. I joined an agronomy group and we employed some good consultants, we met four times a year and toured the paddocks on each other's farms. It was the beginning of the herbicide fallow or chemical ploughing, the start of conservation tillage that eventually led to zero-till, very low disturbance crop and pasture establishment. This seemed to have a lot to recommend it, erosion risk was dramatically reduced due to low disturbance sowing points that left 90 per cent of the soil undisturbed. It was a system that worked, but it was tied to herbicides that made it possible. These were toxic products that were unknown to nature, they had negative effects on soil life, the source of the majority of the mass and diversity of earthly life. After ten years of this type of zero-till farming, I began to question what I was doing.

Our son Matthew, who had been born with a profound congenital cardiac defect, suffered a cranial abscess that threatened his life and we decided to put Allendale on the market and move closer to Sydney and medical care. I enrolled in a sustainable agriculture graduate diploma course and at the same time began a holistic management course. This change of direction was driven by social tension we felt about Matthew's situation, economic tension we felt about increasing indebtedness and the tension we felt about the effect our farming method was having on the earth. In some ways, it was the classic precursor for change that opens the mind to new knowledge and action.

I encountered information in these two academic and practical courses I was studying that expanded my mind. One book that had a big influence on my thinking was Aldo Leopold's classic *A Sand County Almanac*. I read a lot about how the landscape functions, about geological developments over time and the role of living organisms in keeping the world in a life-friendly state.

Fundamental to this philosophical change of heart was the concept of holism. The Zimbabwean-born American-based Allan Savory had written a book, *Holistic Management*, outlining a framework for decision making that had at its heart a holistic goal developed by the people managing the whole (in our case, a farm, but just as applicable for a business, a government or a household). This goal, now referred to as a holistic context, is about: how do you want to live based on your values; what do you need to produce to allow you to live like this (this can be food and fibre, time for recreation, time to allow habitat to be created) and what is the future resource base that has to exist into the long future while improving diversity of life. This three-part goal developed towards what people want ensures that decisions will be simultaneously socially, economically and environmentally sound. In each decision that affects the land, the strengthening of function in the water cycle, the mineral cycle, the capture of solar energy and the dynamics of the living community must be achieved or at least progress must be made towards it. Biological monitoring to track progress towards the goal will reveal if you are moving towards or away from what you say you want. If monitoring reveals you are moving away from your goals it is time to replan and get back on track.

In our case, at Allendale we adopted this different way of decision making just before a nine-year annual rainfall deficiency of between 30 and 50 per cent. Following the principles of what we had learnt we changed our stock numbers up or down to match the available resource, always trying to manage for full ground cover of plant material, living or dead. Our monitoring showed that we started with 70 per cent ground cover, but over three years this quickly rose to 100 per cent or full cover. This eliminated the risk of wind and water erosion, resulted in zero costs for feeding animals (a saving over nine years of between $500,000 and $800,000 compared to what we would have spent in our previous management). We did not increase our debt over this time, compared to those whose crops were failing regularly and who were feeding animals while turning farms into dust bowls and were suffering increasing debt and feeling depressed about how their land was looking. While we were equally concerned at the lack of rain, we felt much calmer than in previous droughts, knowing that even in such trying conditions our land was covered and debt was under control. We also had several catastrophic events to cope with at this time. In 2006 a bushfire burnt one-third of our farm including 23 kilometres of fencing, my father died in October of 2006, the fifth year of nine dry seasons and then, in 2007, our dear son Matthew died of complications to his congenital heart problem. I feel sure if we had been managing conventionally at that time the anxiety of this situation would probably have been too much for us to bear. I am sure we would also have gone out of business and our land would have suffered terribly.

Gradually I began to shift to a form of management that allowed the processes of life to be expressed. We changed our attitude from one of control to one of letting the self-organizing capacity of living communities do the work we once thought was our job. I read a lot of books on ecology and learnt that for complex communities to maintain their structure and diversity they used a substantial amount of solar energy that they fixed for themselves. Our previous farming practices were using up this necessary maintenance energy, which set the community on a trajectory of simplification. Simple communities are the opposite of the trend of evolution and require us to be responsible for many of the processes that complex communities can provide for themselves.

I was single minded in pursuit of ecological knowledge. My perception changed with my understanding. I began to see that many of the costs of my former management style were inevitable as our management produced more and more simplified communities. I had understood through reading that the natural tendency of evolution is to elaborate and diversify the biota and so it made sense to me that we should be farming in a way that allowed this to happen. So we stopped growing crops because the processes in cropping were not leading towards complex communities, which was what we desired. The negative thoughts we were having about our application of toxic products was also far from what we wanted, so we acted by stopping annual cropping over two years and sold the equipment that had large amounts of energy sequestered in tractors and implements. We wanted a farm driven by contemporary energy from the sun, proceeding to a state of complexity guided by the inherent capacity for self-organization within it.

Over the last 19 years we have observed many of the things we desire as people, economically and for the future resource base of the land, beginning to unfold in our paddocks. We have changed from trying to be controllers of the living communities of our farm, to seeing ourselves as part of the living community.

Some of our work has been to stop doing things that conflict with the values we want for ourselves and the landscape we live in. Seeing diversity increasing is good for the human psyche; if you can see it, it has a calming effect.

When we were slaves to the industrial version of agriculture it was as though we had abdicated the process of thought to someone else, the consultants we paid. Now we do our own thinking and make decisions towards the values we want for ourselves, our business and the future landscape we wish to be part of, running on contemporary solar energy, allowing self-organization and renewal to occur and observing diversity increasing over time. It is the living community that holds the knowledge to achieve this, our job is to accept that and allow it to be expressed.

Acknowledgements

In writing this chapter, I have had considerable help from the members of the Fenner Writing Group, members of which have reviewed and made comments in a spirit of critical loyalty that have improved it from its original draft. I especially wish to thank Dr John Harris and Professor Emerita Valerie Brown, who have always been incredibly generous with their time and wisdom. As always I am so grateful to my wife Mary who is always supportive of my various writing projects that need big chunks of time away from other activities. Without her generosity, these projects would not be possible.

References

Carson, R. (1962) *Silent Spring*. Boston, MA: Houghton Mifflin.

Leopold, A. (1949/1989) *A Sand County Almanac and Sketches Here and There*. London and New York: Oxford University Press.

Massy, C. J. (2017) *Call of the Reed Warbler. A New Agriculture – A New Earth*. St Lucia, Queensland: University of Queensland Press.

Rebanks, J. (2015) *The Shepherd's Life*. London: Allen Lane.

Savory, A. and Butterfield, J. (1999) *Holistic Management: A New Framework for Decision Making*. Washington, DC: Island Press.

Part III

The future

17

That's how the light gets in

David Waltner-Toews

At some point in the journey to gather our independent thoughts and make our small contribution to an emergent mind, we may pause and ask why we are doing this.

The world may end tomorrow. If not through wars over oil or water or nuclear accidents or eco-industrial wrist slashing, then through meteors from space or the death of our sun. In just this way, our loved ones may die at any time. Do we therefore love them less? Of course not. It is because of this that we care more. For the tasks of sustainable human communities on this planet, there is, in the words of poet Mary Oliver: 'Only one question: How to love this world.' What does this mean? Send a Hallmark card? Make a flower arrangement? What is it that helps us get up in the morning, humming, perhaps, and actually do something to make the world just a little better?

Romance the earth by candlelight? We could do worse than to look into the eyes of the biosphere and feel, again, or the first time, our primal biophilia. But, in the vast heavy blackness of the space in which we float, from whence comes this light? 'There is a crack in everything', says Canadian poet and songwriter Leonard Cohen in his song 'Anthem'. 'That's how the light gets in.'

To romance the earth, we begin with a candle in the encircling gloom. Some of us have thought of science as a kind of light and it is. The open, mutually correcting community of science, as well as its precise language, so apparently unsuited to poetry and stories, seems suited to the search for exact truth. But such a candle, some of us have discovered, is not enough. The precision of description does not reduce uncertainty of understanding. Neither can it get its linguistic head around the fact that we are inside this earth, having evolved inside it, and have no external objective way of determining whether or not what we think is true actually is. Even such a 'hard science' thing as Bovine Spongiform Encephalopathy (or mad cow disease) eludes us. Here is a poem I wrote about that terrible disease, trying to use the precise language of science, but ending up in the ambiguous mess of real life (Waltner-Toews, 2000):

A bill from the power company
It begins in New Guinea
with veneration of our ancestors.
It begins with love of wisdom and intelligence.
It begins with admiring the dead,
It begins with our cleverness
and our envy.
It begins with eating the brains
of those whom we admire,
with women and children first.
It ends with Kuru, a spongy Jacuzzi
of laid back prions, engulfing the brain.
It ends with a young woman
throwing herself into the fire.

It begins in America with a dust bowl,
with the world war's devastation.
It begins with hungry children
in Europe and Africa.
It begins with our cleverness
and our lust for power and our tractors.

It begins with cows coming in from the green
wilderness in droves.
It begins with praise of hamburgers
on every tongue.
It begins with some spare change
after shopping for food,
and the thrill of a new car.
It begins with recycling, with efficiency,
with cow-eat-cow.
It ends with old men
hungry for power.
It ends with a mob of mad cows
fed to power stations,
shimmering up in smoke
from the incinerators.

It begins with love of life.
It begins in a white coat.
It begins with cutting and sewing.
It begins with new drugs.
It begins with our cleverness
and our fear of death.
It begins with ingesting those
who have what we want.
It begins with blood transfusions,
with hearts, kidneys, and corneal transplants,
with bone marrow and dura mater.
It ends with rabies, with AIDs,
with a slow toppling of prions
across the brain.
It ends with a young man,
unable to walk, unable to speak
his own name.

It begins with flowers and wine.
It begins with clever conversation.

It begins with the love of children,
of making them, and, surprisingly, caring for them.
It begins with a house and a car
and a school and a computer.
It ends with paying
the power bill.
It comes round
to huddling by a campfire
playing old guitars,
singing plaintive melodies,
and, in the ash-black darkness at our backs,
the sound of cows or bears foraging,
and the slow sigh of a rising moon.

So we explore outward into this complex web, beyond the candle and the campfire. Like love, science is curious. Why light, for instance? We all need light. We cannot escape it. We need light if we want to go somewhere, light for our eyes to guide our feet. Still, blind people seem to manage okay, guiding themselves through auditory and tactile clues. For the rest of us – the auditorially and tactiley challenged, however – light is important. We could ask: who needs to go anywhere? We could read books. That is a way of going somewhere without going somewhere. Ah, but the trees that gave their leaves and the rivers or coal or uranium that fed our power plants have sacrificed their light to make our reading possible. Could we sit in the dark and turn white and fat? Many of us do this, with consequences. In the long winter in northern parts of the world people can suffer from seasonal affective disorder (SAD), which is a fancy way of saying that if you are cold and in the dark you tend not to be your cheery old self. The darkness creeps into your bones and into your heart. It settles there like a clammy mould, slowly recycling you for the benefit of bacteria and albino newts. The answer to this, of course, is a big satellite dish with hundreds of channels. But that really is just another form of light. Cold, blue and warped, but light nonetheless.

Even if there were a good comedy channel, we would still need to eat and even potatoes of the non-couch variety, the ones they make chips out of, need above-ground leaves to gather the radiant energy of the sun to come into being.

Some of my ecologist and engineering colleagues speak of exergy – not just any energy, but energy in a useful form. A warm room and the element in an oven may contain the same amount of 'energy', but not the same amount of exergy. You can't for instance, cook a chicken by placing it in a warm room, although you could grow enough bacteria to kill a few people. What we are talking about here is not just a general warm feeling, but about the energy from the sun concentrated and built up into complex webs of plants and animals that we increasingly usurp, starving out other species in our frenzied feeding. This exergy, a kind of gift we have received without asking, we use to build highways and parking lots and black-roofed buildings and plastic garbage bags or refrigerator storage containers or burn to run cars or tanks or airplanes. And yet this scientific view of the light as exergy gives barely an inkling of why we should care. We need that light to live; it is so basic a need that we are willing to steal from other species, scrounge from other people, wrest away from our children, the way we have stolen the forests and the seas and the birdsongs and the terrible marvel of insects from them.

We need more than simple solar energy if we are to care

The darkness is so intense. The newspaper headlines speak of modern-day slavery, national tribalism, ethnic cleansing, the ritual worship of junk-bond traders. Sometimes the moral world we live in seems like such a gloomy place that the darkness is palpable, that morning breaks, literally, shattering in millions of brilliant shards against the unbreakable night. We burn everything we can – trees, coal, electrons – just to hedge ourselves against the omnivorous darkness.

If you get far enough away from the city, away from our electronic fires, out into a field of stubble or at the edge of Lake Huron, or a desert in Australia or Mongolia, you can look out into the vastness of the universe. The darkness, it seems, expands breathlessly in all directions. The stars are tiny points of light, candles flickering in an empty room, speaking to our sense that we are alone, speaking to our hopes.

Or perhaps the darkness is a vast dome punctured by tiny flaws through which a radiance bursts to us. If we saw this radiance in all its glory, we would surely die. Yet, secretly, would we not all prefer to die in such a blaze of glory? When I was growing up, my religious teachers used to talk a lot about conversion experiences. This was a blinding, Damascus Road kind of revelation which would change your life. Some of us felt that we had had such moments.

The trouble is, like any adrenaline rush, they can become addictive and the temptation to try to capture that wild light, to cage it, to make it ours on demand, is almost overwhelming. Like our peers taking psychoactive drugs, some of us got saved many times just to try to keep that life-changing light rushing through our veins. We even carried this with us into our collective political lives. In the 1960s, we spoke of a global, world-changing revolution, ushering in an age of peace, harmony and good music. This sounds more like an epiphany than anything Karl Marx would recognize.

But if the real light, the raw, undomesticated light, cannot be counted on to come in sudden, blinding revelations, neither does it respond well to the flick of a switch. The more we try to confine the power of the Spirit to our religious power stations and spiritual grids, the more that brilliance escapes us. Perhaps our reach should exceed our grasp and we should aspire to be perfect. Yet it is not in the grasp of perfection that the light comes to us. Just when we think we are becoming perfect, when we have achieved the perfect science or the perfect philosophy or the perfect religion, we discover that we are still in darkness. The fundamentals of our lives still elude us. Human perfection is an illusion of perfect darkness. It is in the muddled middle ground, where science and culture meet, where the light comes.

Light strays our way through the cracks

Working in Kathmandu, I was sometimes tempted to see only the filth and disease, the dirty water, the dogs next to the food and the children playing in the garbage. But, as German painter Rainer Nepita was wont to point out to me over pots of sweet, milky Nepali tea, there were also blue and yellow and red plastic baskets piled next to the shiny brass pots, women with the bright saffron and purple saris, earth tones of the street, children, hope, breaks in the clouds. Where I saw darkness, he saw light. Where I saw the broken-ness, the cracks, he saw what was revealed in the broken-ness, the sprouts of green and light.

'Ring the bells that still can ring', says poet Leonard Cohen. 'Forget your perfect offering.' And then, in the recorded version of the song, he does something that I'm not sure even he was aware of. He creates a crack in the line. 'There is a crack, a crack in everything', he sings. 'That's how the light gets in.' The little repetition of crack, separated by a comma, a pause to let in the light. If we want to find the light, we will need to see the cracks.

If there is a crack, that implies that there is a whole which is yet not a whole, something is fragile, but not utterly broken. But in a world seemingly torn between the smug and the shattered where can we look for this light? How can we see this? How can we find this light?

Early one morning after one of my talks with Rainer I went out into the city. When I came back, struggling to cope with the dark chaos around me, I wrote a poem called 'Grace Amid the Ruins' (Waltner-Toews, 2000):

All night
the dog on the roof next door barks.
At dawn, my body's temple echoes
with confused shouts and protests,
brain cells lie toppled,
tangled in ruin.
The hotel guard,
still half asleep
and wrapped in shawls,
releases through barred gates
this fool Canadian,
off for an early morning walk
through night-littered alleys.

Along the roof-line
of Kathmandu
silver satellite dishes are raised
like cupped hands
awaiting crumbs
from the gods:
news from Washington,
Frankfurt,
the Bombay movie scene.

In the asphalt street
a toddler squats

amid the litter of plastic bags,
green mango leaves, shattered red bricks,
squashed yellow-brown mottled bananas.
One of the storekeepers
from across the street
is throwing stones at her.
When she has made
her bile-ish pile,
she deftly drops her tattered hem
and scampers off.

A pre-teen girl
in an earth-tone plain and threadbare dress,
leans like a shaft of slim-limbed light
across a darkened doorway.
A boy, her age-mate, bold
in a clean white shirt
and pressed blue trousers,
self-consciously runs across her line of vision
to the neighbour’s variety shop.
The wry light in her eyes
and the disdainful, one-sided uplift of her lips
tell all: no amount of red and white candy
offered from his hand
can save him.

A slender boy, all ribs and arms and legs,
shifting from foot to foot, shivers
in a bright blue tub, arms folded stoically.
His ten-year-old sister is splashing
water over his bony beige body,
rubbing him down with soap
and rags.

They burst out laughing
at the sight of me.

All day three ragged urchins
worked the crowds,
one rupee, one rupee,
their voices lingered over bright Gore-tex,
tugged at untucked, faded blue cotton.
Now, at dawn, like puppies,
they pile up at the roadside
in sighing sleep.

On the bridge over the Bishnumati
the vendors are laying out
their brass pots, cotton shirts, and fruit,
the tailors are tuning their machines.
Below, the pigs forage in offal
from slaughtered buffaloes
and open latrines.

The satellite dishes bring Bosnia,
Chechnya, Kashmir, Colombo,
the G-7 summit, Seinfeld, Destiny Ridge,
a new world economic order, Mad About You,
beheaded tourists, Friends,
Schwarzenegger, Schwarzkopf.
Men in dark suits
pose for the camera. A voice says
they are making progress.

I look to the hills,
the mountains cloaked in grey rain.
From whence will come our help?
From whence the world's end?
The courage to endure?

The girl on the sidewalk
very carefully stacks her lychees and peaches
in pyramids of five.

With nimble grace and dignity
she crosses her legs
and waits.

But what is this that we all long for and wait for? It seemed to me that if light were grace, which I guess it is, then the girl on the bridge already had grace. More than the bumbling Canadian out for his morning constitutional. But if light is only grace, then there is nothing we can do to find it; grace comes to us, not us to grace. But are there places we can go to where grace can find us more easily? Is there a street corner or coffee shop or a mall where she hangs out?

I once thought that perhaps the girl on the bridge had an answer or that I had one. I now see this isn't going to happen. Neither of us, alone, can find an answer. But together, not just the girl and I, but all of us, might find an answer, might learn our way into a future worth embracing.

Later in his song, Cohen says: 'Every heart to love will come, but like a refugee.' So, we are back to love. After we have tried everything else, this is the refuge we flee to in our leaky boats, in our roped-together rafts of knowledge. Or, as the post-epiphany Saul of Tarsus so eloquently put it: 'I may be able to speak with the languages of men and even of angels, but if I have no love, my speech is no more than a noisy gong or a clanging bell; I may have the gift of inspired preaching; I may have all knowledge and understand all secrets; I may have all the faith needed to move mountains – but if I have no love, I am nothing. I may give away everything I have, and even give up my body to be burned – but if I have no love, this does me no good.'

Having tried everything else – perfected our technologies, our bodies, our minds, done all the self-help stuff, climbed Everest, the spiritual exercises and the bodily exercises, the jogging, the yoga and the eating right – we will, if we still yearn for the light, return to love, flee to her as a last resort.

But are we any further ahead talking about love than about light? Light, grace and love, after all, are all different ways of speaking about something mysterious, powerful and wonderful at the heart of everything. What can we learn from love about grace, about light?

Says philosopher Michael Ignatieff: 'Many of the things we need most deeply in life – love chief among them – do not necessarily bring us happiness. If we need them, it is to go to the depth of our being, to learn as much of ourselves as we can stand, to be reconciled to what we find in ourselves and in those around us' (Ignatieff, 1984).

And what do we find in the depths of our being?

Here is a riddle. In loving one another, in loving this wonderful biosphere of which we are a part, we acknowledge our separateness even as we desire wholeness. Unless we are each alone, individual, whole in ourselves, there is no one to summon love into being. Unless we are incomplete, fragments of some larger whole, we will not yearn for love. For love to become, we must be whole and we must be broken or at least cracked.

It is tempting to believe that this is all just about the fluttering of our hearts, about the light that breaks through the gloom of a winter's day when we unexpectedly see that special someone we care about, recklessly and anxiously, about teenagers trying to see the shape of one another in the darkness in the back seat of a car. But this is not just about us and our friends, about the pain of saying goodbye and the hope of seeing each other again.

This goes to the very core of ourselves as humans. Our brains are cleft in two. The left half of our brain, so some researchers have said, is more analytical and logical; the right half is creative and intuitive. What makes us who we are is not just the two pieces, but the communication across them: the corpus callosum, those bundles of nerve cells, traversing the clammy, dark and uncertain depths of the longitudinal fissure. They connect the two halves of the brain. That's how the light gets in.

This goes, as well, to the very heart of our political and cultural being. It is about the most powerful and destructive forces unleashed in the past century: the fierce desire to make us all one and all alike, as in communism or Nazism and their ideological and religious offspring; and the opposite, the nativist capitalism, tribalism and libertarianism, based on the unshakable belief that individuals are everything. Both are a refusal to accept a fundamental contradiction of life.

Ignatieff, in writing about the 18th-century philosopher Jean-Jacques Rousseau, says the following: '[His] insight is that a community of men can become masters of their needs, instead of slaves to their desires, only when they democratically decide upon some form of collective constraint on inequalities of fortune ... Apparently, societies that seek to give everyone the same chance at freedom can only do so at some cost to freedom itself... Modern secular humanism is empty if it supposes that the human good is without internal contradiction' (Ignatieff, 1984).

There is a crack in everything

This goes, as well, to the core of the great questions of environmental sustainability we ask today, of the possibility, which many now seriously doubt, that human communities can flourish on this planet without destroying the incredible diversity of other communities of living things that make those human communities not only liveable in, but possible.

Biodiversity has become a watchword for environmental activists around the world. But biodiversity is not just about all the pieces, about saving germ plasm in banks. That misses the point entirely. More than anything, biodiversity is about relationships, what goes on in the cracks between things. The wonder of nature is not just the cardinals, squirrels, the beautiful quetzal birds and lemurs and white-tailed deer and timber wolves. It is not just the riches of life we find recorded in the archaeological records. It is the Wagnerian chorus of them all together, the tongue-tickling harmonies, the tum-tiddling down your vertebrae, the humming in the stem cells in your bone marrow, the twisting in the ancient codes that make us who we are: the Auruch, the Quagga, Gypsonia, Dickinsonia, Glyptodont, the Moa, the Ammonites, the Trilobites. It is about the chorus of history and evolution bursting through us, poetry of the Burgess Shale, the Java Man, my grandmother, the Canadian Valkyrie elisions of people and grizzly bears and raccoons and trumpeter swans and even elected politicians.

This is not just about us, but about the families of which we are part and the communities made up of our families and the ecosystems made up of families of millions of species. This is about being a whole and being incomplete. This is a love, then, that requires an all-encompassing care, an aching tension and contradiction and struggle. Perfect equilibrium in any living thing only comes with death. According to thermodynamic laws, we must live in tension, far from equilibrium, or we cannot live at all.

The philosopher and novelist Arthur Koestler created a language that I think can help us here. He coined the word 'holon' to describe something that is both a part of something and whole in itself. A holon, being whole, is internally rich with relationships, like all the psychophysiological pathways that make you who you are. But it is also a part of something larger, communicating, sending out and receiving essential messages to the larger holons, the family, the community of which it is a part. A holarchy is what a holon is part of. We are done with hierarchies, with bosses and boss-ees, with one-way, downward communications. As we must rediscover, the fundamental rule of the biosphere is not based on authority, one over another, but on serving one another, not as shuffling slaves, but as holons. We love our neighbours – and our enemies – as we love ourselves. We communicate with one another. We converse. We exchange gifts. We sing. We write poetry and do science. These are the relationships that make our collective life on this planet possible – and wonderful – and it is the collective life that makes our individual lives possible.

This then is our challenge: to be ourselves as individuals, and at the same time to claim our part in the whole of this wonderful creation. To be holons, simultaneously acknowledging our separateness and our wholeness. To find ways for people of many religions and cultures to be true to themselves and yet not to demand that everyone be like them. To love our neighbours, our enemies, our fellow species, as we love ourselves, as unique individuals, as whole, and as part. This is the contradiction at the heart of everything, the crack where the light gets in, the challenge for a new kind of science and a new kind of storytelling, a new, integrated way of thinking about knowledge, a post-normal science, a science where the peer group is expanded to include a large motley crew of our fellow citizens, and where 'hard science' meets cultural storytelling.

If we keep our senses alert and our mental eyes wide open, moving like the half-wild savannah animals we are, we may find the skills necessary to be good citizens of earth as well as good scientists, to see, and, seeing, as Pablo Neruda said, to come to life, to see 'the fire that sprang to life in beautiful things'. We will then find those places that the light and grace will come to. If we are willing to be alienated, to be lonely and to be compassionate, then we will find the light that comes through the cracks. Maybe not a blinding flash of insight, but enough to travel by, enough to read by, enough to see where we are, the wonder of it and to make our way.

References

Ignatieff, M. (1984) *The Needs of Strangers*. New York: Penguin Books.

Koestler, A. (1978) *Janus: A Summing Up*. Melbourne: Hutshinson.

Waltner-Toews, D. (2000) *The Fat Lady Struck Dumb*. London, Ontario: Brick Books.

18

Knowing our own minds

David Waltner-Toews, Valerie A. Brown and John A. Harris

David

The great Baltic German Biologist Jakob Johann Baron von Uexkül, who laid the basis for what came to be known as biosemiotics, probed the ways in which non-human animals perceived, and made sense of, their *Umwelts*, the worlds they perceived and in which they acted. In the century since he died, we have discovered the hidden communications of insects and trees. We have even, in the words of science writer Ed Yong, exploring the microbiomes of the world, discovered that each of us 'contains multitudes'.

No less than all those other species, each of us humans emerges from, and lives in, our own *Umwelt*. We are, each of us, within ourselves, manifold: physical, social, ethical, aesthetic, sympathetic. How we feel and what we think, what we believe is right and how we act are often in anguished conflict.

Given this, how can we think independently, and then act in a world where every shift in behaviour, every exploration, nudges the whole system in new and often unexpected directions? How do we take responsibility for individual decisions that have collective consequences and vice versa? The authors in this book have asked those kinds of question and investigated them in theory, empirical research, debates and arguments.

This book is part of a conversation for which we have yet to find a common language, a shared grammar. Or perhaps that is wrong. We do not seek, like the engineers and building contractors in the myth of the Tower of Babel or the 17th-century Latin-educated European scientists to find one language. We are struggling to find ways to listen to, and to interpret, voices across many languages.

Many indigenous tribes of the northwest coast of North America have a tradition of the 'talking stick' or 'speaker's staff', which is passed from person to person in a group, allowing each to speak and to be heard without interruption. In the spirit of this book, and that conversation, we will each hold the talking stick for a few moments and then present a collective offering for consideration.

Valerie

The future of the world is uncertain, not predetermined. I am hoping it will be whatever emerges from people around the world thinking independently and interactively. That future will surprise and amaze us all. Independent and interactive thinking, when set in place, time and history with other such minds, will generate new ways of engaging with and understanding the world, transcending what went before. The problem is, as we found in the research reported in Part 1, not all individuals have access to their own independent thinking.

Every epoch has its own mind. We have had the Renaissance and the Reformation; the Anthropocene should be no less mind changing. In this, my own epoch, I will need to be both independent and interactive, hopeful and realistic at the same time. While as individuals, we editors may disagree with one another, the spark between us opens up other ways of thinking. The writer-naturalist, Aldo Leopold, is an optimist, based on his vision of love for all life. Buddhist economist Clive Hamilton is a pessimist, despairing of humans' ability to learn from lessons from the past. How can we bring them together? Kate Andrews has described regular swings from optimism to pessimism as a circular conundrum; a continual cycle of hope and despair unless and until an independent mind breaks the cycle (Chapter 10).

It is a challenge to hold two opposing ideas in my mind at the same time. The normal conclusion is that one idea is right or at least more right than the other. On the other hand, as an independent thinker I will consider whether both can be right. A crucial 'and' joining optimism and pessimism brings me a realistic hope of a new direction, unpolluted by the dominant thinking that separates them. Philosopher of ideas Isaiah Berlin presents a stark account of the overall situation for humanity, not as alternatives, rather as one confusing world. He lists moral blindness, egoism, cruelty, humiliation, servility, poverty, helplessness, bitter indignation and despair, on the one hand, and truth, love, honesty, justice, security, dignity, decency, independence, freedom, and spiritual fulfilment, on the other. The point he makes is that each of us must make our own way through this mental jungle. I will need to learn how to hold contradictory ideas in my own head, rather than blindly following the usual socially generated pathways.

A way of thinking that breaks the cycle of hope and despair, while allowing for both, has been called the 'emergent mind'. An emergent mind is always emergent, never expects to be final and never closes itself off from its society. It is open ended, able to critically examine the society that shaped it, and which it shapes in return. Such a mind builds on the strength of the multiple dimensions of an individual mind explored in Chapters 1, 2 and 3. The originator of the idea of the emergent mind, writer-farmer

Charles Massy, discusses the implications for a new agriculture that he labelled regenerative. Regenerative describes a process of renewal, building on the past and hopeful for the future. Since agriculture is the foundation of human existence, Massy argues that a new agriculture gives birth to a new world (Chapter 9).

Having described himself a regenerative farmer, David Marsh then describes his farming practice as not intervening in a natural system or imposing a social system. Rather he 'allows' each system to develop to the fullest use of its resources, giving his support when needed (Chapter 16). Massy reviewed over 90 successful regenerative farmers in Australia, Africa and the United States and, in 2018, Marsh was awarded the National Landcarer award. From this we can assume that regenerative thinking is an emergent idea. Massy and Marsh report greater resilience to droughts and floods, security of income, time for families and friends and personal satisfaction (Chapters 9 and 16).

The idea of regenerative practice can be applied to human social systems. Different cultures, occupations and ages each develop their own version of reality, the collective coherence described by consultant Craig Ashhurst (Chapter 13). As a regenerative thinker, I would respect others' worlds and support others to do the best they know how. A necessary step for a harmonious future is for us to reconcile, not to avoid, dissident voices. The normal response to reconciliation in our epoch is to try for consensus. However, consensus is regarded by political scientists as the most conservative of responses: sometimes it can lead to the lowest common denominator and sometimes to the most powerful individual winning the prize. Post-normal thinkers use current thinking as a point of departure for fresh ideas, not as an established basis for solving their existing problems (Chapter 1). Professor of indigenous health, Kerry Arabena, applies her Torres Strait Islander sense of the whole to regenerative practice. She argues for a post-normal reconciliation between members of different cultures, ages, religions, and regions, grounded in multifaceted postmodern science (Chapter 7).

An emergent mind, regenerative practice and post-normal reconciliation that take account of all the dimensions of mind are developing just below the radar. They threaten and alarm the 'normal' sources of power, who move to silence them. However, individuals have their own voice and it is time our voices are heard. In Chapter 6, John Harris and I developed a vocabulary for a possible new language. This post-normal language takes account of both connections and divisions, when the normal practice has been to emphasize the divisions. Lawyer and environmental activist Dianne Dibley's search for an environmental ethic brings together the knowledge compartments of economics, science, ethics and law for individuals from the silos to work together on a shared ethic (Chapter 8).

Nowhere is this post-normal thinking more powerful than when we draw on the multiple dimensions of our minds. This possibility, of minds being divided into physical *and* social *and* ethical *and* aesthetic *and* sympathetic – *and* to be whole at the same time – is at the core of independent thinking. I have written elsewhere about the power of replacing 'or' and 'but' with 'and' and 'also' (Chapter 3). When using 'and', then as Leonardo da Vinci insisted, everything is connected to everything else. For Charles Darwin, five years spent on the *Beagle*, and for Nelson Mandela, 27 years spent in gaol, gave them time to reflect on all dimensions of their experiences. Their writing reveals that they drew on the whole of their minds to give us holistic understandings of evolution and of freedom. Their reflections gave them access to all the dimensions of their minds; to a new view of the world around them, sensitivity to their society, strongly held ethical positions, creativity in breaking the mould and acute sympathy for others.

The change to multidimensional and holistic thinking requires a dismantling of the current hierarchy of knowledge and power, and its redistribution to access for everyone (Chapter 2). In this process, it is critical that young people, especially, engage with life and politics so that democracy itself becomes regenerative. The post-normal agenda requires emergent minds to come together in open collaboration and dialogue, making a distinction between 'I' and 'thou' (Chapter 15). A redistribution of power means giving a voice to the powerless. That, in turn, requires reconciliation between the powerful and the powerless, with all that that entails. Change agent Julia Wolfson describes this as her professional practice in working with both carers and the cared-for in human services organizations (Chapter 14).

It often seems to me beyond my capacity to resolve the complexity around me. Yet it is part of the miracle of human thinking that I respond to this complexity every day. What I do not do often enough is to reflect on my own thinking. Luckily many independent thinkers have left me a legacy. At the close of Chapter 6, John and I develop criteria by which we can recognize our own independent thinking. First, because we are human, we each have the capacity to think for ourselves. Second, we try to build on the tensions inherent in diversity, not try to remove them. Third, our mind remains open to any possible (and seemingly impossible) alternatives. Fourth, we redefine the normal version of the issues to be resolved, not accept them as given. Consider the locked-in positions on climate change and the denial of the obesity epidemic. When I practise these ways of acting, I can go beyond existing locked-in positions. The outcome is regenerative in the fullest sense of the word.

Even as we speak and write these words, the languages of our conversations are changing. In the 1970s gender equality came into the general consciousness. Among the tensions this caused was the recognition of a gender-oriented language. *Man's* place in nature, chair*men* of committees, prizewinning sports*men* and the use of a generic *he* excluded women from life, power, competition and even existence. The simple changes to human: chair, sportspeople and his and hers; this simple difference was enough to change minds and hence actions. I am old enough to remember the hysterical laughter and tightened lips of members of an audience that first heard the use of 'she' throughout a scientific paper. A similar shift in language supported the post-war peace movement: objectives, not targets; storing resources, not filling war chests; and let's talk, not fire away. In our multidimensional minds we can use language to draw on all physical, social, ethical, aesthetic and sympathetic facets of our minds, and not restrict our language to only one category of mind.

In Table 18.1, the multiple dimensions of the minds of individuals and of society are matched with the emergent ideas that encompass both.

Table 18.1 Individual, society and multidimensional thinking

Dimension	*Individual*	*Society*	*Multidimensional*
Biophysical	Empirical	History	Evolution
Social	Identity	Ethos	Social contract
Ethical	Personal	Cultural	Spirituality, grace
Aesthetic	Own feeling	Social rules	Symbols
Sympathy	For one's own	For others	Empathy

Figure 18.1 **Head of Janus, Vatican Museum, Rome. Photographer: Loudon Dodd. (source: Wikimedia Commons)**

As in the case of Hercules, discussed in Chapter 1, complexity of thinking and feeling and acting is deeply rooted in our mythologies. The Roman god Janus faces both ways: endings and beginnings, past and future, beauty and ugliness, war and peace, open and closed, right and wrong and individual and society. The separation of these pairs opens up a theme pursued throughout this book, the need for 'and' in an uncertain world. For wise decision making, we need strong individuals in a strong society (see Figure 18.1).

John

Learning to become a skilled independent multidimensional thinker requires a supportive society in which individuals are free to attend some form of open-minded public education. In my long teaching experience, when young people are asked to engage in real issues, they naturally use all of their mind. At the same time, their ever-active brain establishes the complementary neural pathways and networks for fitting it better to their developing independent mind.

The paradox that then faces such strong individuals is how to work collaboratively as full members of their society. Evidence for this occurring on a small scale is provided throughout the book, but to scale this up in a society is not easy. Yet the challenge of living in an uncertain world demands nothing less. My hope is for diverse multifaceted approaches to grow in societies worldwide through the evolution of individual and collective emergent minds. The possibility then exists for a global field of shared meaning and positive energy to flourish among all peoples of the world.

My intuition is that multiple forms of dialogue offer a fruitful pathway for independent thinkers to engage positively in society. David Bohm found listening without judgment vital (Chapters 5 and 15), David Kolb used experiential learning cycles (Chapter 2). There is Open Space Technology's 'this is the right time', Julia Wolfson's facilitation hubs (Chapter 14), the talking stick referred to by David and my own field study classes (Chapter 5). Successful dialogues exemplify the active experience of individuals listening to and respecting one another; suspending their judgement before speaking their own voices. Bohm's genius was to realize that multiple points of view must be expressed before opening up new possibilities and multiple options. It is this flow of shared meaning, often unspoken, that holds groups and societies together and provides the fertile ground from which individual and collective thought emerges.

The writing group in this book required skilful facilitation and the participation of all its members to create a generative dialogue. In essence, two kinds of condition must be met in any society. These are the quality of 'collectivity' of society itself and the independence of strong individuals within society. Once these twin conditions are established, it is not long before things become self-organizing, allowing the creative processes of individuals and groups to emerge. A synergy is then possible between our species' propensity for learning and the use of our multidimensional minds. Emergent properties are generated through collaborative learning, independent thinking and openness to change. Independent thinking can then flourish in collaborative groups throughout society and independent-minded individuals act as catalysts for change.

I have experienced this in different working contexts over my lifetime. I have noticed that individual creativity flourishes best when individuals allocate time for reflection on the physical, social, ethical, aesthetic and sympathetic dimensions of their thinking and being. This opens their minds to seeing the 'bigger picture'. And this, in turn, generates a vision of the possibility of contributing to a larger world beyond their own.

David

Perhaps it comes from putting together too many Powerpoint presentations or to compensate for the endlessly restless snakes' nest of neural connections that seem to characterize my thinking, but I will summarize in point form what I am taking away from work described in this book.

Independent thinking is relational, that is, like the biodiversity in natural systems, it emerges within and from particular, diverse human and natural ecologies and histories but, by being part of these systems, nudges them in new directions.

This particularity of thinking within place, time and history, set into relationship with other such minds, generates, through those relationships, new ways of considering and engaging with and understanding the world that transcend those particularities.

The visionary, global, noospheric mindfulness envisioned by scholars, investigators and thinkers of the past century is emerging from the necessary tensions and feedbacks and debates between independent thinkers.

Like most trends in natural and human ecology, both negative and positive, this dynamic is cumulative and can lead to surprising, often unexpected, tipping points.

The work ahead is to stimulate independent, mindful, relational thinking. Karl Popper (of all people, since he is viewed as a stalwart of scientific reductionism) has talked about the propensities in nature that carry events in a particular direction. In this Anthropocene age, we are almost certainly part of that momentum.

I see this happening in the kinds of project we describe in this book and, like the dramatic social and political changes that have occurred in the past century, these events are accumulating just below the radar, occasionally threatening and alarming those in power and often minimized by them, gaining momentum decade by decade.

At night, at the edge of sleep, my mind grapples with all the contradictions and excitement that the conversations in this book (and, indeed, in my life) evoke. Listening to my mind debriefing, I wonder what it means and am left in awe, in wonderment:

A Chat Among the Neurons

The words unravel
guilelessly,
a thread pulled here,
a lip unbuttoned there.
When eyes ask questions
that the tongue cannot,
the world's undone.

The mind sings
in polyphonies,

and glossolalic birdsongs,
conjuring children, weather,
the spread of plagues,
of jam on toast, of legs
a good idea has
to change the world
one moment – this moment –
at a time.

We speak of energy,
vanishing whales and jaguars,
of reproductive fitness,
of how the meaning changes
under the weight of all
our bodies, and the last refuge
is invaded, and we are
awe-struck by the mountains,
by a glimpse of what's uncovered
by receding glaciers,
unfound species, specious
arguments, until
but one thought's left,
the only one that matters,

and all night I lie eyes-wide,
my auditory nerves
patched into the callosal commissure,
monitoring flashes of traffic
to and from Broca's hillside estate,
mysterious electric zings
and will-o'-wisps, faint babelings
of pre-hominid tongues,
straining to recall
what *is*, after all,

and what it is
that really matters.

Together, if we listen in mutual respect, if we help one another find the words, we can discover what it is that really matters.

A conclusion

With an emergent mind, the treasure is an ideal toward which we travel. Who knows what will happen on the way? We are like the travellers in Chaucer's *Canterbury Tales* and in B. P. Cavafy's poem *Ithaca*. The great value of the destination – of the treasure – is the journey it offers us. Let us hope, as Cavafy advises, that 'the journey is long' and, unafraid of the Laistrygonians and Cyclops we carry within us and the 'wild Poseidon' who attacks us, let us care for one another on this journey. Let us be mindful of the wonders around us and of one another, listening and conversing, telling each other stories, as we walk.

Some books offer dreams, visions and theories. Others demand our attention with 'hard evidence' and documented, corroborated experience. In this book, we are asking you, and us, to embrace all of these. Philosopher Karen Houle, in an exercise with participants in a course on ecosystems and health, asked us to hold as many contradictory thoughts and feelings within ourselves for as long as we could without exploding. This, she said, was complexity. We, the editors and authors of this book, are asking that of you, and of us. The future is uncertain and what will emerge in that dissilient, exploding, emergent mind when people around the world think independently and interactively cannot be known, but will most assuredly surprise and amaze us all.

Bibliography

Arabena, K. (2015) *Becoming Indigenous to the Universe: Reflections on Living Systems, Indigeneity and Citizenship.* North Melbourne: Australian Scholarly.

Berlin, I. (1959/1990) *The Crooked Timber of Humanity: Chapters in the History of Ideas*, Hardy, H. (ed.), 2nd edn. Princeton: New Jersey Press.

Berlin, I. (1998) *The Proper Study of Mankind: An Anthology of Essays*. London: Pimlico.

Bohm, D. (1994) *Thought as a System*. London and New York: Routledge.

Bohm, D. (1996) *On Dialogue*. New York: Routledge Classics.

Bonneuil, C. and Fressoz, J.-B. (2016) *The Shock of the Anthropocene*. London: Verso.

Darwin, C. (2015) *The Voyage of the Beagle: The Illustrated Edition of Charles Darwin's Travel Memoir and the Field Journal*. London: Zenith Press.

Hamilton, C. (2017) *Defiant Earth: The Fate of Humans in the Anthropocene*. Crow's Nest: Allen & Unwin.

Kolb, D. A. (1984) *Experiential Learning: Experience as a Source of Learning and Development*. Englewood Cliffs: Prentice-Hall.

Leopold, A. (1949/1989) *A Sand County Almanac and Sketches Here and There*. London and New York: Oxford University Press.

Mandela, N. (1994) *Long Walk to Freedom*. London: Little, Brown & Company.

Marsh, D. (2018) https://www.youtube.com/watch?v=3VoeMDFdTQk [accessed 11/5/18].

Massy, C. J. (2017) *Call of the Reed Warbler: A New Agriculture – A New Earth*. St Lucia: University of Queensland Press.

Wolfson, J. (2017) *Applying Deep Democracy in Human Services: Diversity, Inclusion and Innate Powers*. San Francisco: Eldership Academy Press.

Index

For Product Safety Concerns and Information please contact our EU
representative GPSR@taylorandfrancis.com
Taylor & Francis Verlag GmbH, Kaufingerstraße 24, 80331 München, Germany

www.ingramcontent.com/pod-product-compliance
Ingram Content Group UK Ltd.
Pitfield, Milton Keynes, MK11 3LW, UK
UKHW021504240926
475702UK00035B/360